ALL BED AND WORK

LOOKING AT LIVES OF LANCASHIRE TEXTILE WORKERS: BURNLEY 1975

CHRISTINE SLATER

TO THE FOND MEMORY

OF JIM SLATER

A PILLAR OF THE TEXTILE INDUSTRY

TABLE OF CONTENTS

PREFACE

Women textile workers have long formed a special category of industrial workers and have been at the forefront of women's industrial employment. Yet Lancashire's cotton mill weavers and spinners, especially women, have remained relatively unknown figures.

The stereotypes have been there right enough for a century and more - of clogs and shawls, of clean provident and house-proud workers, with scoured front steps and black-leaded stoves.

On the other hand there have been stories of 'gormless' male loom over lookers and of independent, bold girls with loose morals and indecent language.

Now even the stereotypes have become become dim, as mills have closed down decades ago and remnants of mill life are seldom seen, other than in local museums.

The evidence for this book was collected nearly forty years ago, in the mid nineteen seventies, when the mills and their culture were still to some extent vibrant, though rapidly changing.

What was it like then in the seventies at the close of an era? It was a time when the cotton industry was undergoing another transition before its virtual demise.

Married women still formed the backbone of the labour force.

The mill girls of all ages no longer wore clogs and shawls. Indeed they had not done so for fifty or more years by then.

They were more likely to wear mini skirts and platform soles.

What did they think about themselves and their jobs? How did they combine their several roles as workers, wives, housewives, kin and mothers, and possibly good neighbours and autonomous, glamour girls?

How did men and women relate to each other on the shed floor and in the home? What did they worry about? What did they gossip about? In the words of a weaver, what made them tick?

Some answers to these questions were sought during several weeks of listening to cotton workers - men and women, young and old – talking about themselves, their work and their home life.

Some of the things they said are presented at length in chapters 2 and 3. In chapter 4 are comments and descriptions by one or more of their own work mates.

These are based on notes written over a period of time by my late mother, Mary Slater, who was a mill worker for several years of her life - both in her youth and middle age.

She was first briefly a weaver, when she left school and then much later she taught weaving.

In these observations of a participant in mill life and work, we are given some lively insights into circulating gossip.

We hear snatches of dialogue between work mates, female and male, young and old, locals and immigrants.

We see some of the kinds of domestic and personal problems they have on their minds, as they try to start a new job and the things they are concerned about.

Mary never had a chance to go to secondary school or college.

But in her work in the mill she was able to use her undoubted, albeit untutored, skills as counselor and teacher and in her observations showed her promise as a budding ethnographer!

The first chapter is on the ups and downs in the cotton industry itself. It draws heavily on several of the works listed in the bibliography. It is meant to form a brief historical back cloth to the subsequent chapters. It gives a summary account of the industry over time, depicting the kinds of changes which took place, affecting the employment prospects and conditions of the workers, both men and women.

Times were often tough or very tough.

Work was usually grueling and hard.

It was also for the most part in the past poorly paid and insecure. Stories and memories of unemployment and destitution were only too vivid among the elderly.

But people were highly skilled, - women as well as men - and took great pride in jobs well done.

They lived through periods of great industrial output and success, making great profits for the mill owners and survived with great struggles through the periods of downturn and unemployment.

The final chapter consists simply of scraps of conversations overheard in a mill. These give vivid impressions of the way t h e w o r k e r s talked and some of the things they talked about.

They provide a unique window and an insight into some of their life values showing, amongst other things, their great sense of humor. Stories and conversations are recorded and presented as they were heard and some contain bad language and sexual references, which have not been omitted or censored. No political correctness here! Words are used which would not be countenanced in polite society today!

By the end t h e r eader is, we hope, left with some vivid insights into the lives of an important section of the British industrial community, which is no more.

All names of people and factories have been changed, as have some place names, to preserve the privacy and anonymity of all the people who so readily took part, many of whom are now long gone.

An attempt h a s been made to replicate the typical Lancashire dialect used by the workers.

The text has been written without end notes or footnotes for simplicity and ease of reading.

A bibliography provides the sources of the historical, factual material about the industry.

A short epilogue and end note, based on materials from some available websites and recent publications, highlights some developments.

There could be a growing resurgence of interest in the past lives of the British working class, particularly in the light of some of the more derogatory, - nay even demonizing, - views aired of late in some quarters about "chavs"!

This account is ultimately meant to provide more insight and greater understanding of textile workers' lives, in Lancashire in the early, middle and later decades of the twentieth century and how they were rapidly changing at that time.

Christine Slater January 2013.

ACKNOWLEDGEMENT

The materials presented in this book were collected in the heart of the Lancashire cotton district, Burnley, in the mid nineteen seventies. I was joined in this task by several factory friends and relatives. Without them this work would never have been started or completed. The collabouration and help of all the weavers, winders, spinners and over lookers, who gave their time and talked so freely and enthusiastically is gratefully acknowledged. My main inspiration and guide was my father James, who introduced me to the workers, who told their stories. This text is dedicated to his blessed memory. My key collabourator was my late, beloved mother, Mary.

Christine Slater 2013

THE AUTHOR

The author mainly grew up near factories in Lancashire textile towns and villages - Ribchester, Longridge, Blackburn and Burnley. Her father had to leave school at 14 to help support his family. His own father had been badly injured in the First World War. He keenly felt the deprivation of higher education. He was first a weaver, then over looker, then a self taught, weaving shed manager. He was deeply respected, loved and admired by his workers.

The author's mother and mothers' sisters, from a farm family, were all sent to learn to weave in Ribchester, as soon as they left school. The author was one of the fortunate daughters of textile workers who were supported through higher education, by parents dedicated to the goal of learning. She benefitted from High School education in Blackburn and Burnley, where she was Head Girl, and then won a State Scholarship, to fund a first degree and a State Studentship to support   further graduate studies in social anthropology.

1.   COTTON: ITS UPS AND DOWNS

The cotton industry was one of the first major industries in the world to use power driven machines and employ large numbers of workers in factories. The cotton fibre was shipped east to the United Kingdom from the West Indies and North America, through the port of Liverpool, to be woven in Lancashire and elsewhere in Britain. The skills of weaving and spinning had been developed in a thriving cottage industry, originally using wool and flax. Already by the early nineteenth century cotton weaving had become Britain's biggest manufacturing industry and North East Lancashire produced the bulk of the plain cloth.

One thousand two hundred of the one thousand six hundred mills were located in Lancashire. Blackburn, Burnley and Preston were the biggest weaving towns. Bolton, Oldham and Rochdale were the main spinning communities.

As early as the eighteen twenties Blackburn was producing nearly 50,000 pieces of calico a week and Burnley and district 25,000 to 30,000. Manchester was the commercial centre of the industry.

In the beginning some of the biggest hand loom manufacturers employed even more workers than did the factory owners.

One firm alone is reputed to have put out work to thousands of weavers in North East Lancashire.

Many of the hand loom factories had up to twenty or more looms and a few had up to a hundred or more. But the typical hand loom weaver worked on their own looms in their own house, using materials from a manufacturer or their agent, who paid them according to the cloth woven.

There were cotton workers in the mills in the nineteen seventies whose grandparents remembered seeing their fathers and uncles weaving cloth at home and carrying it to be sold. The heyday of the hand loom weavers spanned the period 1770 till 1840. Some villages and towns in North East Lancashire were very heavily populated with weavers. Nearly all the houses in Great Harwood for example were said to have had looms and figures from some of the small villages in the Burnley district suggest that half of their inhabitants were hand loom weavers.

It was common to combine hand weaving with a small holding, to grow vegetables and keep livestock or to also ply another trade.

From the start women formed a large proportion of these skilled workers.

As early as 1808 half of the weavers were women and children and by 1833 women were in the majority among the calico weavers in the Burnley, Colne and Skipton areas.

During 1820-1850 the number of power looms as opposed to hand looms increased rapidly. These looms, which became known as Lancashire looms, were semi-automatic. They were invented by James Bullough and William Kenworthy. They were designed such that when a warp thread broke, the weaver was alerted and when the shuttle became empty the machine stopped working, so weavers could manage four looms or more instead of one. It wasn't until 1900 that the Northrop fully automatic loom was developed, letting a weaver manage more looms, but that was suited to coarse thread not fine.

By 1835 Burnley had 10 small mills with 1,165 power looms between them.

The first factories in Burnley were built on the banks of the river Calder, near to the juncture with the River Brun, as water power drove the spinning machines.

In addition local coal mines provided power and the Leeds Liverpool Canal was important for transport of bulk goods, while local foundries and iron works provided the needed machinery.

Smaller towns and villages such as Colne, Clitheroe, Accrington and Padiham had one or two factories each.

In contrast by then Blackburn had 3,200 power looms.

Bury had 9,000 and Manchester itself had 15,000.

The first half of the nineteenth century was a period of increasing hardship for the hand loom weavers, who suffered periods of widespread shortage of work and consequent poverty and hardship. Their resentment against the innovation of semi-automatic power looms burst forth in the loom breaking in 1826.

Since the weavers were concentrated in fairly limited areas, whole communities suffered en masse.

During this period, even with full employment, a weaver could scarcely feed their family, as the power looms forced the piece rates down. Local charities were unable to cope with the size of the problem. For instance in Blackburn in 1826 more than half of the town had to be relieved weekly with food for nine weeks to prevent widespread starvation and death.

Memories passed on of such events continued to haunt people, when work was difficult to find and available help meager or non-existent.

The census figures of 1821 to 51 show shifts of people from more remote villages into the towns, where factory employment was situated, but the main demand was for girls and women.

It was difficult for men to find factory work.

By 1850 cotton hand loom weavers were rare and by the 1860s there were only a few.

By the 1880s it was reckoned that there might have been three or four working in the villages of the Ribble Valley -around Longridge and Ribchester.

From 1750 till 1850 cotton hand loom weaving had provided employment for thousands of people - many of them women - for the next hundred years the power spinning and weaving looms did the same. By 1878 Burnley for instance had 80 firms. By 1902 there were 87 weaving mills, 2 spinning mills and 18 mixed.

So fast did the weaving industry grow in Burnley that in 1886 the town boasted that her 50,000 looms produced a greater length of cloth than any other town in the world.

Overhead charges - rates, gas, coal and water were low and Burnley workers had a reputation for skill, efficiency and thrift. In addition excellent mill engines and plain looms were manufactured locally.

Burnley's population growth from 4,840 in 1801 to 93,920 in 1897 indicated the thriving expansion of the town during the period.

Eventually it boasted of becoming one of the largest, if not the largest, producer of cotton cloth in the world. The number of looms installed increased rapidly from 2,600 in 1841 to 55,136 in 1897.

In 1848 there were 25 firms engaged in cotton manufacture in the town.

By 1897 this figure had grown to 110.

In the period of rapid expansion of trade, speculators built sheds.

Looms could be bought on hire purchase and local families seized opportunities to start businesses. The canal which was opened in 1806 introduced cheaper transport facilities and mills were erected on the banks of the canal, which also provided a ready supply of water for the industries needs. The discovery of coal in the middle of the town had played an important part in the development of textiles and mining developed considerably during 1860- 1870. Working conditions remained extremely poor and often dangerous and hours worked by children, women and men were incredibly long and grueling. It was not until as late as 1902 that the minimum working age was legally raised to twelve years. Till then cruel child labour had been a heinous blot on the industry.

The Peak of Properity

A high peak of prosperity in the textile industry was reached in 1912- 14.

This was the period which those retired in the mid seventies remembered their early years as tenters (starting as learner weavers aged twelve to fourteen).

Lancashire cloth was clothing the peoples of the world. Exports of seven thousand million square yards per year accounted for nearly two thirds of the total cotton piece goods then traded in the world.

In addition over eleven hundred million square yards a year were woven for the local market.

The industry had led the world in the industrial revolution and was then supplying the major part of the world's requirements of cotton cloth.

There were 61 million spindles, 786,000 looms and 620,000 operatives.

In Manchester, the market place for the sale of cloth and yarn, there were special groups of merchants to deal with the Indian and West African trades.

The Inter War Years 1914-1940

After the First World War, there was a short boom of high prices and high levels of profit for mill owners. The apparent need for greater output attracted more capital and more looms and spindles were installed, but this extra capacity was soon found to be in excess of the real demand, and by 1920 competition from abroad was felt and exports dropped.

By 1921 prices had fallen and there was a recession. In 1914 40% of exports had been to the Indian market but in 1914-18 there was a period of mill expansion in India and by the nineteen twenties exports to India had fallen to half of their pre-war level.

The effect of the diminishing of the Indian market on the British industry between 1912-1938 accounted for a drop in one third of the size of the British industry.

In addition there was Japanese competition in far Eastern markets and European competition in near Eastern markets.

Thus by 1926-8 Britain's percentage of the world cotton trade had fallen from 59% to 39%.

The home market was gradually expanding but it could not offset this massive decline in exports. Exports fell from 7,000 million square yards in 1913 to 4,500 million in 1920, to 1,716 million in 1931 and then down to 1,386 million in 1938.

From 1922 onwards the number of looms installed declined rapidly. The nineteen twenties and thirties saw a number of mergers and attempts to reduce the surplus capacity of the machinery and also to cut down on the number of operatives.

In 1932 the suggestion that weavers should run more looms was put forward by employers, to help increase output and reduce costs.

Weavers were to run eight looms, with help for sweeping, cleaning and carrying cloth.

Burnley was the town chosen to give the *more looms* system a trial for twelve months.

The local weavers' association opposed the scheme.

The outcome was a strike in which eleven mills stopped work and eventually all except one small mill were closed.

The situation was grim for the striking weavers and a fund was set up to prevent starvation. Those in greatest need were given half -crown food parcels.

Six mills reopened and employed weavers from neighbouring towns, who were out of work.

A large crowd of out-of-work weavers demonstrated against this.

A police man was injured and arrests were made. Many people got kicked and bruised.

There were violent demonstrations outside mills which were opened. The police had great difficulty in maintaining law and order.

A procession on its way to Brierfield was stopped by mounted police from the Manchester city force. The procession was going to Nelson to agitate for a strike of all cotton mills. During this time there was much hardship and suffering, ill feeling and allegations of police brutality.

Towards the end of August, when Burnley workers had been on strike for five weeks, a general strike began. With the involvement of the whole cotton industry, the Ministry of Labour decided to intervene and set up a special committee.

Ultimately the *more looms* agreement was signed and only sixty to seventy percent of the old workers were re- employed. During the period the Local Education Authority had provided more free meals for schoolchildren.

A distress fund had been set up and distributed over a thousand food parcels a week.

During the strike the Co-operative Society had also given two hundred parcels a week but the relief had been insufficient. And over the next few years more than six percent of the population left to find work elsewhere.

The history of Burnley town mirrors the changes taking place in the cotton industry over time.

In 1920 there were about a hundred firms in the district, operating 100,000 looms and nearly half a million spindles.

From 1921 onwards with the slump came the closure of many mills.

Some looms were exported, others were scrapped. There were three months in 1921 when not a single Burnley mill was working. By 1929 the working, insured population was estimated to be about 49,000. Among these there were approximately 31,000 textile workers.

Many endured long periods of unemployment and, as the industry declined, thousands left. In 1929 cotton accounted for 63% of the industry in Burnley, but by 1938 it had dropped to 43% and unemployment was a bane. One result of the persistent depression in the area was the stream of migrants to other areas. Between 1930 and 35 the population declined by 7,000.

The depression of the textile industry itself was deepened by the great general business depression, which affected all countries and industries after 1930.

In 1935 the Burnley Borough Council tried to improve the situation by encouraging other industries. Old cotton sheds were reconditioned and offered to new industries. By then a third of the weaving firms had ceased to exist.

A development committee was set up with the object of establishing new industries, such as the production of kitchen utensils, handbags, furnishings, leather ware etc. in order to promote stability of employment in the town.

To improve the general situation of the industry a Spindles board, set up in 1936, assisted in the disposal of spindles.

And looms were disposed of mainly by individual firms.

Accordingly by 1938 only 7% of installed spindles and looms were actually working.

Another government reaction to the drop in exports was to enforce special protective measures to support cotton exports to the Dominions and Colonies, so that by 1938 - approximately three quarters of cotton goods exported were protected in some way.

And by 1939 the quantity of cloth exported was less than that bought at home. Exports were by then only half of the 1930 level. During the second world war the industry was concentrated. Labour and space had to be contributed to the war effort and cotton supplies dwindled.

About a third of the capacity of the spinning and weaving mills was closed down.

Production, employment levels and exports all fell drastically. In the meantime the cut back of British exports was paralleled by an increase in production of mills abroad in Latin America, India and the United States .

After 1945

After the war from 1946 till 1951 labour was scarce and had to be attracted. There was growing competition from other industries.

The chance to recapture the world markets was lost.

This was partly because the industry was unable to recruit enough workers to achieve the necessary levels of production.

So difficult was the labour situation that by 1948 1.5 million spindles and 30,000 looms were still lying idle in closed mills and from 1948 onwards the machinery in these closed mills was scrapped at an increasing rate.

In addition the modernization and re-spacing of some machines to improve conditions of work, led to some reduction in capacity of many mills. By 1947 the industry was still producing less than half its pre-war level.

From two thirds, the value of textile exports had fallen to one tenth of the value of UK exports.

Increasingly raw cotton was being woven by the countries which produced it.

From 1949 to 53 the loss of exports was partly due to restrictions on imports into India and Pakistan and competition from India and Europe.

After 1953 the competition came mainly from Japan and Europe. From 1955 exports were contracting and textile imports rising.

By 1952 a recession had set in and the contraction of the industry continued. Some firms closed down. Many mills went out of existence or were taken over.

In spring 1952 the drop in output was drastic. It fell to half the 1951 level.

Between 1952 - 4 eight thousand looms a year and eight million spindles were cut down.

This was achieved partly through re-spacing and replacement of machinery and partly through mills closing down.

In 1953 imports were restricted.

The industry had reached 90% of its 1951 peak level and could not meet the demand for cloth. In 1954 importation of cloth from India and Hong Kong into the home market led to a new recession. This in turn led to a reduction in capacity of at least 20%.

Workers were leaving the industry at the rate of a thousand a week.

In 1952 they left at the rate of 2,000 a week. Between 1951 and 1966 about 1,000 mills closed down and output dropped by 45%.

The number of jobs fell by 60%. By 1967 the number of mills and finishing works still in operation was less than 750.

Reorganization & Modernization

An important criticism of the cotton industry has been its inability to modernize and innovate and adopt new technical developments.

To an increasing extent innovation in textile machinery occurred abroad, in the United States, Japan, France, Germany and Switzerland.

The rate of re-equipment with modern machinery was extremely slow.

Information was made available, but it would seem as if the will to change was lacking.

Admittedly the advantage of the industry's early start in the nineteenth century became a disadvantage in the twentieth century.

For the industry was left with a lot of old, outdated machinery.

In contrast in the new textile industries, developing in other countries, modem equipment was bought.

There had been so little change that by the beginning of the second world war the organization of the industry was very like what it had been just before the first world war, although it had run down.

No new mills had been built since 1925 and few since 1914.

Trade was declining. The labour force and the machinery were both becoming older.

It still remained the greatest manufacturing industry in the country however, and an important contributor to the country's exports.

Modernization was a slow process. In 1948 only 7% of looms were automatic and by 1958 it had only increased to 14%, leaving a lot of outdated machinery still running.

Many of the weaving mills had been developed as small family concerns and still in 1958, of 900 weaving firms the most typical unit was running fewer than 100 looms. Two thirds of the companies owned less than 500 looms each.

On the other hand 25 firms with over 2,000 looms each accounted for a fifth of total capacity.

Similarly five spinning firms controlled a third of the spindles, while 390 units were controlled by over 200 firms.

The trend however was towards the concentration of production into fewer manufacturing units. Many companies merged, reduced their capacity or closed down. In 1953-4 alone 95 companies merged or closed. In 1959 the Cotton Industry Act was passed.

The aim of this act was to bring about the contraction and reorganization of the cotton industry, by subsidizing the scrapping of obsolete plant and assisting re-equipment of the remaining mills.

The government bore two thirds of the cost of buying up machinery to be scrapped. There were fixed rates of compensation.

Firms were given two months to decide what to do in August and September 1959.

Employees as well as employers were paid compensation and the scheme was administered by the cotton board.

Subsequently there were re-equipment schemes for the installation of new machinery.

The five hundred and nineteen weaving firms had an average size of 103 employees and 22 of these left the industry.

Many of the small mills in remote villages continued in existence however, depending for labour on the local married women.

Of the machinery 38% of the looms were scrapped and 8% of the spindles.

Thus there were marked changes in the size and structure of the industry between 1959 and 64.

By 1965, 250 small weaving firms had left with a mean size of 130 employees.

The 1959 Act was aimed at improving the efficiency of the existing industry.

However after 1964 the changes which occurred were much more radical in nature.

Under pressure from imports the decline in the industry continued.

In 1960-8 the industry was radically changed by a number of mergers and take overs.

In the take over process the main six buyers may have bought something like 100 to 150 companies.

An important restricting factor in the changes taking place in the industry was the age and immobility of the labour force.

Married women preferred to continue doing part time or shift time work at the local factory, to full time work at the more distant mill. They were highly skilled and they were not mobile.

Thus during the 1950s and 60s a massive upheaval took place in the cotton industry, with the concentration of production in fewer units and the concentration of ownership in fewer hands.

The industry became smaller but more modern and m o r e efficient.

This period also saw the development of a number of major innovations in methods of textile production and processing.

These included the development of new synthetic fibres, new methods of chemical treatment and the development of high speed machinery.

A major change was the replacement of the old Lancashire looms.

In these each new length of weft wound round a pern which had to be inserted into the shuttle by hand.

In contrast there were various types of automatic looms, in which the shuttle refilled automatically with weft.

By 1961 out of 159,000 cotton and synthetic fibre looms installed in Great Britain 48,000 (30%) were fully automatic and 2% had been fitted with automatic attachments.

Another development was the shift from cotton to man- made fibres.

Competition from synthetic yarns had increased gradually. By 1951 a tenth of spinning was rayon yarn and a quarter of woven cloth was rayon mixtures.

The synthetic fibre revolution was destined to bring about fundamental changes in the structure of the entire textile industry.

At the same time the market situation radically altered. In spite of the reorganization and modernization however, many of the very small plants remained in existence.

But they were uneconomic and unable to afford the cost of re-equipment with modern machinery.

And the see saw, up and down character of the Lancashire textile industry persisted for several decades, except for a few short periods of relative stability.

Indeed it still continued into the mid- seventies, when the stories of Burnley textile workers told here, were collected.

The Workers

A picture of the industry has been briefly sketched above.

In particular we have indicated how it changed over time during the twentieth century.

The question remains, who composed the work force, from the early nineteen hundreds to the decade of the seventies?

How did they survive and adapt to the deprivations and fluctuations in prosperity that it brought about?

How did they and their families cope?

By the mid nineteenth century there were already more than half a million workers in the cotton industry, and the majority were women.

Evidence from as early as 1839 had showed that over 56% of workers in cotton were women. Many were the wives and daughters of men working in the mills.

Women soon earned a reputation for being better suited to weaving than men, being noted for their speed, skill and deftness.

In weaving women were included in the unions from the start and both sexes were paid at the same rate of wages. But women's wages in textiles as a whole were lower than those of men.

This was because their hours were generally shorter and men rose to become over lookers, foremen and shed managers, on higher rates of pay.

The peak size of the textile labour force was reached in 1912, when there were 622,000 workers.

From then on a long decline set in, till by 1967 there were only 120,000.

By 1921, in Lancashire as a whole, the proportion of metal workers was higher than the proportion of male textile workers.

But for women, textiles was still overwhelmingly the main occupation, with 380 out of every 1,000 occupied females working in cotton.

In many of the smaller towns and villages the majority were cotton workers.

The spinning area was by then concentrated in the south east of Lancashire in Oldham and Bolton. The weaving districts concentrated around Blackburn and Burnley.

The area of most intensive concentration spread 22 miles from north to south and 32 miles east to west in East Central Lancashire.

The Burnley district with the highest concentration of male weavers included Accrington, Trawden, Colne, Barrowford, Nelson, Brierfield and Padiham.

The Blackburn district included Chorley, Darwin, Long - ridge, Clitheroe, Great Harwood, Clayton le Moors, Rishton, Preston and Haslingdon.

In between the spinning and weaving areas was a mixed belt, including towns such as Rochdale and Bury.

Weaving was largely a woman's occupation.

Indeed in the neighbourhood of Burnley, there were communities in which nearly all working women were employed in cotton factories, mainly as weavers.

For example in Trawden 90.5% of employed women were in cotton; Great Harwood 86.9%; Padiham 84.8%; Brierfield 83.6%; Oswaldtwistle 83.3%. In Trawden 43% of all women and girls aged twelve and over were textile workers, mainly weavers.

A marked difference between women textile workers and other women in factory employment then and later on, was that they did not stop work at marriage or even child birth.

The majority continued to work during the years of peak domestic responsibilities from 20 to 45 years.

The proportion of married women in the occupation was noted in 1921 to be higher than in any other occupation. They also outnumbered men in textiles overall.

In Burnley in 1921 there were 11,734 male textile workers, of whom 7,445 were weavers, while women textile workers numbered 20,217, of whom 16,327 were weavers.

Between 1918 and 1940 short time work and unemployment were widespread.

The figures on unemployment in the Burnley area between 1929 and 1936 give some indication of the situation of insecurity and anxiety born by the workers.

Between ten and thirty five percent were unemployed in any one year.

In Great Harwood between 1930 and 31 the percent of

unemployed reached 70 and in many areas the unemployment rate was higher among women than men.

Unemployment reached such a peak that in 1931 out of the 1,169,100 employed in textiles a total of 220,600 were out of work - almost nineteen percent.

This included 49,800 cotton spinners out of work (22.4% of the total) and 96,800 cotton weavers (28% of the work force). In other words one in five spinners were jobless and nearly one in three weavers.

The steady dwindling in the size of the industry led to a halving of the work force between 1914 and 1938.

Then again between 1931 and 1951 the number of women occupied as weavers decreased by 55% and winders etc. decreased by 29%. Nevertheless women still formed 65% of the working population.

During these years of contraction and unemployment, the industry was unattractive to the young school leavers.

The trend towards an ageing labour force was strengthened by the subsequent war time measures, which affected the younger age groups.

Thus the average age of operatives gradually rose from 29 in 1911 to 31 in 1924 to 34 in 1931 to 37 in 1946.

This trend continued, as other avenues for employment of the young increased.

In the twenty years up to 1951 the proportion of weavers over forty five doubled, rising to 38%.

By 1952 while nearly forty percent of male workers in all industries were under 35, among cotton weavers and other workers the percentage was twenty two.

The majority of the latter were in fact over 45 (63%) as compared with 37% of all workers. Indeed almost one in ten of male cotton weavers were by then over 65.

As already noted one major response to the lack of secure working conditions in the cotton industry was for people to migrate elsewhere in search of greater security. And this hundreds of people did, not only to the Midlands or South of England, but also abroad and so to this day many Lancashire people have relatives in Australia, New Zealand, Canada and the United States.

The result of this migration was a marked dwindling of the size of the cotton towns from 1911 onwards.

Since men formed the majority of the migrants another result of the movement was that Lancashire between 1921 and 1951 was one of the counties with the highest proportion of females. This trend towards depopulation continued in the Blackburn and Burnley districts.

In addition many children of mill workers got an education and moved out to regions of greater opportunity and prosperity and alternative employment.

Between 1951 and 1971 the population of Burnley further decreased by over 8,000.

By 1951 a total of 877,000 were employed in textiles and only 1 percent were unemployed.

The industry was booming again. The demand for more workers remaining unsatisfied and many different kinds of cloth were being woven, from corset cloth to car upholstery.

In 1951 however only 29% of the labour force in the area was in cotton and there were 1,500 mill vacancies at the local offices of the Ministry of Labour.

Women were still playing a major part in the work force and over forty percent of the working population were women, a proportion nearly ten percent above the average for the country.

At that time the four towns with the highest percent of women in the work force were Oldham 40.3%; Rochdale 40.5%; Burnley 40.9% and Blackburn 41.9%.

But in 1951 -3 memories of the pre-war, short-time working and unemployment were reawakened.

The industry was once more regarded as lacking stability of employment. When it revived only a third of those leaving came back. Sixty thousand workers left in six months.

Some married women left and stayed at home or were absorbed into other industries.

Between 1952 and 5 the loss of labour was rapid and substantial.

Many experienced workers were entirely lost to cotton. In addition the recruitment of the young was interrupted.

At the low point of the recession a fifth of the labour force left.

Subsequently the trend of production remained downward, as the industry was less able to cope with competition at home and abroad after the loss of labour.

Ironically by 1958 lack of labour supply was an important limiting factor in the industry and by then many operatives were in the older age groups. The average age of workers was 40 in 1954-8.

The labour force had been further reduced by nearly 78,000. There were four million fewer spindles and 60,000 fewer looms.

Even during the early sixties there was a marked drop in the numbers of textile workers.

The number of men in textiles dropped from 168,080 to 161,130 and of women from 297,69 to 243,580.

In the case of married women the proportion working in textiles as opposed to other jobs dropped dramatically from 442 per 10,000 economically active to 293.

By 1961 there were 252,370 women textile workers and only 146,70 men.

The largest age group for women was 55-9 and for men 45-54. Several solutions to the labour problem were applied.

Flexible working arrangements were offered to experienced, married women.

For they could often work part-time at odd hours

to fit in with their family commitments - such as mornings or a late evening shift.

More efficient training programs were instituted to train new workers. Better canteen, sports and travel facilities were introduced to attract workers and the doors were opened wider to immigrants, who on some shifts began to form the majority of workers.

These were mainly from India and Pakistan. During the nineteen sixties there was a sizeable influx of Asians. By 1971 there were over a thousand residents from the new Commonwealth, including over 700 Indians and Pakistanis in Burnley alone.

This number was small however in comparison with a number of neighbouring cotton towns.

Thus by the time of the 1971 census there were over four thousand Indian and Pakistani immigrants in Blackburn and Bolton and over three thousand in Preston and Rochdale and Oldham.

These immigrant workers formed an increasingly large proportion of the workers in the cotton mills.

The problem of attracting a stable, skilled workforce remained pressing

In 1974 a mill union official was cited in the press as saying that the textile industry was in danger of grinding to a halt, through lack of workers.

In fact production was only being maintained through employment of immigrants and the employment of women and young people on an overtime basis.

The problem of recruiting a new labour force was growing rather than diminishing. So much for the overall growth and decline of the industry and its work force and the fluctuations in production and employment prospects.

The concern from now on in this work is with individuals in the mid seventies.

We focus on these men and women and what it meant for them to grow up and work in these cotton towns and mills during the preceding decades.

Memories stretch back to the peak of prosperity in 1914, through the years of depression in the thirties, through two world wars and to the later fifties boom and subsequent reorganization and decline.

We see why people went to work in the mills with their continuous thundering din and dust, rather than other jobs. We see how the conditions and people changed over the decades.

We look at what some of the women a n d m e n were like, who from the beginning played such an important part in this major British industry.

We begin our glimpses of the cotton characters with the oldest - those who by the mid seventies were already retired. They had begun work at the beginning of the twentieth century. They could remember many ups and downs and changes. Cotton had been their life.

2. THE RETIRED: COTTON WAS THEIR LIFE

In the mid seventies, the time when these stories were recorded, the population of retired textile workers was comparatively large and many still continued to work full or part time, after reaching sixty or sixty five. In 1966 there were 38,000 retired women textile workers, including nearly 20,000 over seventy. These women were among those who had struggled through the years of depression and unemployment in the 1920s and 30s.

They started at the beginning of the twentieth century, in the era of half timers, the age of clogs and shawls and Saturday morning work, when over-lookers could be bullies, managers autocrats and mill owners dictators ; - when faults were fined - when hot water for tea was paid for and canteen and toilet facilities were few and far between.

Working conditions were poor- looms were close together- wages low and hours long. Many of the workers' houses were built back to back and in the shadow of the mill walls and chimneys. Single streets contained several mills.

The air was never free of their soot and smoke.

In cases where the mill owner also owned the houses, then his control of the workers was even more absolute.

All Bed and Work

In this chapter three women, one man and two couples tell something of what it was like for them. Frequently husband and wife worked in the mill, even the same one, like Clarence and Ethel. For them cotton was their life in a very real sense. Work at the machines was labourious and long.

There was little time and energy left for other activities and then not only whole families, but whole villages and towns, depended upon the production and sale of cotton cloth for their living. Many aspects of home life and work are revealed in these accounts - the strict discipline, the widespread deprivation, the almost universal lack of choice, the pervasive insecurity and frequent authoritarianism. But at the same time the strength and warmth of ties with kith and kin is seen; the tremendous self-sacrifice and ambition of parents for their children; the self restraint and devotion of couples; the common solidarity of kin and the frequent intermeshing of personal ties in the shed and the home.

At the time they started work there was little choice of occupation in the cotton communities of Blackburn and Burnley district. In the several towns 70, 80 even 90 percent of women and girls were employed in the textile industry.

Even for men the percentage reached over 50% in towns, such as Nelson and Trawden.

Thus for the majority of school leavers, work in the mill was a forgone conclusion, as it was for the people who tell their tales here.

Whole families were employed in the mills, in some cases the girls were weavers and winders and the boys went down the pit.

Often parental pressure was exerted, if a child wanted to do another job which did not pay as well.

Some children became accustomed to the clatter of the looms while still at school and even fascinated by the quick moving machinery and cloth before they were considered old enough to be employed.

As Dan says in his youth, "there were only th'pit and th'mill". Those like Ethel who badly wanted to do something else more pleasant but less well paid, such as working in a flower shop, could be dissuaded by their mothers, - desperately in need of money to raise the family.

There was little labour turnover or changing jobs, unless through being laid off or "sacked".

For as Maggie says, once you got in the mill you had to stick it out and the people at work were nearly all the same lot. Mill conditions are indicated.

Gertrude's description vividly tells what working in a hot, dirty spinning mill was like.

There were no places for changing, washing or eating facilities in the factories – hot water for tea, toilets, oiling of looms had to be paid for at 1d or 2d per week out of already meager wages.

In addition fines for poor work were common. A continual source of difficulty and inconvenience was the close spacing of the looms with only 8 - 9 inches in between.

Anybody fat or heavily pregnant could not get past and it was particularly back breaking for the over lookers, when they were carrying the heavy warps on their shoulders.

The changes in working conditions which gradually came were much appreciated, such as the 7am instead of 6 am start; the stopping of Saturday morning work and holidays with pay. Gradually too the wages increased.

Styles of hair and dress also changed in the mill like every where else and while everyone used to wear clogs and shawls, by the end of the first world war, shawls had gone out of fashion.

Before that they were worn by mature girls of sixteen or more who had three looms. Aprons too were different for girls and women.

The custom of tacklers (over lookers) wearing white fustian trousers and aprons gradually disappeared.

Make up was uncommon and hair had to be tied up or plaited for safety, near the spinning belts and wheels. Early to bed and early to rise, was the rule with paid knockers to bang on the bedroom windows to prevent sound sleepers from being late for work.

Women were perhaps under the greatest burden of strain. They had the responsibility of household chores and children, as well as factory work, without the benefit modern machinery, for as Maggie says,

"It were a damned hard life for a woman an' they all had their homes to run and there were no vacuum cleaners or anything like that".

Women were, as Clarence and Ethel agreed, not only the backbone of the cotton industry but the backbone of their husbands as well.

Working mothers usually either got a relative to care for their babies or took them to a neighbour's from six o'clock in the morning till six o'clock at night.

Some stopped off work while their children were small and took in washing and other people's children. Some babies were brought to be fed at work. Other mothers dashed out at breakfast time to feed them.

Some like Ethel sent the child away all week and brought them home only at the weekend.

When work was hard to get men walked or got on their bikes and cycled for miles in search of a job.

Waiting for work meant showing your face every morning at the mill wearing clean clogs and white fent in the hope that the factory foreman might pick you for a day's work.

Every year a dreaded time for family men was the payless summer holidays, when some walked or cycled to nearby towns to get a bit of casual work.

The unemployment and financial insecurity caused couples to put off their weddings for lack of accommodation and money.

Dan and Jane courted for ten years because they couldn't get any "backing" .

No doubt the same factor was influential in there being so many small families of one or two children among the weavers.

Those with the smaller families were the most fortunate financially as they had the fewest mouths to feed.

Material deprivation and physical hardship are recurrent themes of these accounts,- suffering both in the house and at the mill.

Living space was cramped and privacy almost non-existent for families living in one up and down or two up and down houses, where bathing and washing had to take place in the same room as cooking, eating and living.

Food was at the best plain and wholesome and home baked - potato pies and potato cakes, rice pudding, black puddings, tripe and trotters, bread and dripping. At worst it was inadequate and lacking in basic nutrients.

Workers set off to the mill on bread and treacle like Clarence and came home to chips without fish. Children were lucky to share an egg. An important area of deprivation was higher education.

Boys and girls who passed the secondary school entrance examination were not usually allowed by their parents to take up their places in the grammar schools.

They had to start work instead and supplement family income.

As Dan said, parents wanted their children's wages in those days. And it was a case of children having to get down to work.

When working they were lucky if they got a penny in the shilling of their wages to spend or the odd coppers.

Poignant memories, of little luxuries enjoyed or not obtained, illustrate the precarious level of living, such as stockings to match a costume or to go dancing, or ribbons to wear in the park.

Such deprivations were most severe during the thirties.

That was a time when self respecting men found themselves rooting in the gutters for cigarette ends.

Careful housewives had to beg housekeeping money from their better off relatives.

And there were whole years when couples could never leave the house for a night or afternoon out for lack of money.

Men, women and children did all kinds of jobs to earn an extra shilling, from cleaning flues and taking in washing and children, to helping farmers and keeping poultry.

Life did have a lighter side however, both in the mill and outside.

In the factory there were the annual parties at Christmas, even if they were not allowed to switch off the looms and later coach trips and dinners and sports clubs.

The savings banks helped workers to save for these entertainments.

Some at least could afford these things, especially in middle age.

Meanwhile practical jokes and the odd smoke and giggle in the toilets broke up the working monotony for the young.

In leisure hours favorite pastimes were penny pictures, football and dancing.

The church too provided many welcome social activities for church goers.

Family ties were important inside the mill as well as out - for getting a set of looms, for getting promotion and for being admitted into the over-lookers' union.

Young couples often lived with kin, so solving the problem of lack of accommodation and baby-minding.

Thus Dan and Jane lived a year with her brother after marriage and then seven years with her widowed mother. Fanny too lived with her mother.

Mothers, sisters and aunts living nearby could provide essential help in looking after children of working mothers and borrowing of money and goods helped to bridge the income gaps.

Often relatives at work taught the new child how to weave. Fanny's brother taught her, Adam's sister, Maggie's mother.

And relatives, husbands and wives worked in the same shed even loom to loom.

One thing appreciated by any worker who experienced it was to have a manager or mill owner, they could talk to and tell their problems.

Some owners however, obviously maintained a great distance between themselves and their workers.

As Clarence remarked, they looked at the weavers like dirt, ordering them here and there.

Some workers such as Maggie's father were also treated as private servants.

Many tacklers had reputations for bullying and meanness, for favoring the weavers they liked and ill treating others. Connections in the mill were essential to get a job, to keep it and to get promotion.

As Jane says when jobs were so scarce you needed someone to speak for you, someone to plead.

The over-lookers' union in particular was a closed shop. For they had a reputation of not wanting to admit outsiders. Their own sons stood the best chance or friends' sons.

The weavers often felt they were on top of them because in many cases the over- lookers' wage depended upon the output of his weavers' looms and if a weaver's wage was poor he might show his disapproval to the weaver by throwing her wage at her at the end of the week.

So much in awe of tacklers (over lookers) might the weavers be, that they would not dare to lift up their heads when they walked past.

The older weavers in their turn could be awkward and strict with the young learner weavers and might show their disapproval by kicking their ankles with their metal tipped clogs.

The retired s a w clearly the yawning gap between their own generation and the young people in the seventies.

They saw the easy money coming in - £32 instead of 32/- shillings and being spent as easily.

They saw the a p p a r e n t l y   i n c r e a s i n g lack of discipline at home and at work and the lack of pride in the job, which they valued.

They perceived it as a shift from craftsmanship to machine -minding and they knew the problem was not lack of jobs but lack of workers in the cotton industry.

As Clarence summed it all up, in their day they thought about work, now they think about money. It was "*all bed and work* " at the beginning of the century. In the seventies it had become, "*a piece of cake - with a terrible lot of easy money*"!

In spite of the grueling working conditions, conducive to hernia, d e a f n e s s, bronchitis, emphysema and other troubles, as well as the insecurity and low pay, the general consensus was that mill life was hard but happy.

They were even sorry that they had to retire.

Dan would have gone on Saturday afternoons and Sundays i f it had been possible.

He enjoyed every minute of his working life and it took him six months to recover from his retirement. Similarly Gertrude, Jane and Maggie all said that they would go back if they were younger.

Clarence summed up the situation of the cotton factory work by saying that working in the mill is like a virus that catches you.

The lone dissenting voice was Ethel's. She would never like to live her life again.

She would never go back and c o u l d n ot think why they did not have the sense to leave.

She also stressed the mentally deadening effect of the continuous din and motion of the looms.

Although they say they would themselves go back, for the most part they nearly all had other ambitions for their own children.

The common attitude was that of Dan and Jane, who felt they sacrificed so that their daughter could work in an office and not have to cope with the grim life they suffered.

The pride of those whose children went on to higher education is illustrated by Clarence and Ethel, who struggled to send their only daughter to a fee-paying school and to university.

Clarence was disappointed when she became a part time secretary and not a journalist, as he had fervently hoped.

Gertrude Snape

Gertrude's pride and joy is her new council house, which she keeps spick and span. In the parlor is a green, plush suite to match the leaves sprouting on the wall paper.

Her favorite pastime, when she's not decorating the house, is doing jig-saw puzzles. She would go back to work she says but her legs won't let her.

My mother was a weaver and my father worked down th' pit and mi sister worked i' th' mill. I never knew much of her after she got married. Mi' husband were a weaver till about two years ago. Now 'e works at an electrical place.

I used to live down Pit street for fourteen years an' as time went on it got slummier and slummier, and we were told they had to come down. Well, I didn't want to leave, I'd been happy there.

It were near to town, and everything, but they were pullin' 'em down. I landed up here near th' park but I didn't like it at first in this council flat. I couldn't get set in it but now I wouldn't like to go back on a street. And it's very convenient for the buses.

But what makes it worse is that I don't like paying the rent. It's nearly £11 for a fortnight. Well, you see I was only paying rates where I was before. It was my own home. I got paid out. They were going to buy the house off me. I can't grumble at that.

We waited six weeks for the notice to come. Then they come and told us they had to be pulled down. They said they'd put me in something decent if I'd wait, any way I did get something decent. I think I've been lucky.

Mi' dad's mother were a winder. (I never knew mi' mother's mother.) She worked while she were seventy. She'd a very big family. She'd twenty two children. One died. Her husband died. Then she got married again and come to have another child. She always said she'd twenty two children twice. I lived wi' mi' grandmother more than what I live at home. I remember we used to go to th' pawn brokers every Monday morning. Mi grandmother used to collect all their clothes up, and I used to go with her and drag a truck and go to the pawn brokers.

She used to collect for th' neighbours as well. They used to give her so much for taking them, and then when she come out she'd say, "You stick hold o' that and I'll go and have a drink, ' am dry".

An' sometimes I'd be there while turning out time, sat into the truck waiting on her.

Then we used to go home right over th' fields and she wouldn't have any of her own money, and t' other people' d be waiting for their money.

An' sometimes she wouldn't have theirs either. Many a time when I've gone wi' her like that o' th' Monday, I've had to wheel her back home.

I used to love going to mi' grandma's. She used to take washin' in. She used to wash for four families and they were all colliers. She used to bake for people an' do all their patchin' and mendin'. She looked after fourteen grandchildren.

We'd a big long kitchen, wi' three tables put together and a form at each side, an' th' boys sat on one side and all th' girls sat on th' other.

Grandad used to make broth on th' fire, potato pies and crumpet cakes. He used to stand at th' top of th' table and we used to have to send all th' plates down to him and he'd serve it out.

He got fed up with us all saying we didn't want this and

we didn't like that, so one day he piled all the food together in one big dish in th' middle of th' table an' it looked so awful it made us all feel sick and we couldn't eat it.

I used to go round wi' 'im wi' a donkey and cart, and go to all these shops .... Grocers' shops for fruit and that for th' pigs. I 'ad to sort it out an' any that were good I'd to wash it for makin' pies an' the bad was used for feeding th' pigs.

I wouldn't eat it 'cause I knew where it had been, I daren't tell th' others where it had come from. Any cabbage he collected he'd take th' middle out, and make broth, and the rest went for the pigs an' that's how we were kept.

We were well fed. We always washed it well. We had a big boiler in a building where he used to cook his pig swill and we used to love goin' in there.

He used to kill th' pigs at Christmas an' we used to have some good times. I'll tell you.

All th' boys in them twenty two children went down the pit. I remember. Two got killed in th' First War.

Some of 'em went to Durham to th' pits. Some of th' girls were winders in th' mill. Aye she were a strong woman my grandmother. She were eighty nine when she died.

I don't know much about mi' husband's relatives at all. I remember one time my uncle had been cleaning his chimney and he'd been up into th' cock loft to do it. I followed him up on th' steps and saw 'im take two bricks out o' th' wall put one of th' bricks into a sack, and put it down th' chimney.

But I'd never seen what he'd done down at th' bottom. Anyway, a few weeks after that our chimney were smokin' and everything were black, so I said to dad,
"I'll see to it", so he said, "Right, you go up, and I'll stop down here",

So I goes up into th' cock loft, takes these two bricks out, puts one in a sack on a rope, I shouts out, "*Are you alright?*"

An' I thought I heard him say, "*Yes*", so I drops the brick down the chimney. He let out a yell an' ran down stairs an' I could n't see a thing for black soot. He was covered with it from top to bottom. I got the belt and put to bed.

We didn't get much spending money an' everybody went to th' pictures on Monday nights an' we couldn't go. Mi' mum said, "*If you can't save your bit of spending money then you can't go*".

Anyway we decided to go up th' street to th' cemetery collecting jam jars. We got two carpet bags full, then took 'em to th' shop to collect a farthing **for th'** little ones and half penny for the big ones.

He wanted to know where we'd got them from but we wouldn't tell him, so he sent for us parents. We had taken them off each of the grave stones where they had flowers in, and they made us take 'em all back again and put th' flowers back in. So we didn't get to th' pictures after all.

I went to work half-time at twelve and I worked half time for twelve months.

I started as a doffer in a spinning mill. I'd to carry a tin on mi' head and work in bare feet.

It was hot sometimes it used to be up to 90 °F all day. We didn't use to work with hardly anything on. We'd just a smock over us because carrying tins an' like ...you'd to work with your hands putting one off, and putting one on, on a thousand spindles and that carried on all day.

And after you'd done so long at that you'd to try and learn spinning. Well when I was fourteen I went learning spinning and I carried on with that for about six months and then they gave me two sides to run.

It were very rough ... dawny ... dirty really. You used to come out with your hair all full of dawn .... your hands all greasy. We 'ad nowhere to wash.

For drinking water you'd just a little fountain. Well you'd nearly all of you going over it.

Really in them days compared with now it were absolutely filthy. We used to strip off up at th' top end o' th' alley. There used to be little cubby holes an' we used to put all us clothes up there.

We were always full of dawn ya know. We used to take us shoes and stockings off, and when it got hotter we used to take us jumpers off.

We had one of them old fashioned smocks on.

We were always sweltering when we were coming home. Us clothes were full of dawn.

They never made anywhere to put anything we'd just to hang 'em up on a nail.

I can't say anybody were bad wi' me 'cause I always did mi' work.

We used to have to take us own bobbins upstairs and I were bending in this big skip for one and fell in the basket.

My mates just tied it up and I were there for four hours! Nobody 'od tell where I were, and they were looking up and down all over for me. I kept shouting and shouting and nobody heard me, as nobody worked round that end. Eventually I heard footsteps and I shouted out for all I were worth. It were th' manager. He cut th' rope and give me a good telling off, then started to laugh. He brought th' three others up, what 'ad done it, and said what 'ad happen if they did it again.

These that had been on it and that was on the spinning frames, - you'd to go round with them and learn it, - like another person.

And they used to see how you was going on, and if they thought you wasn't fit for it, well you'd come off and you'd to go back on your old job, and they'd put you on again, -

say in another few weeks time.

Well after you'd learnt that, you see you got from one frame to one and a half, and then you worked up to four frames, and they wasn't well paid for it.

I think I'd only about £2 odd and I started at 4/- for half time; 8/- when I went full time. When I went on my first frame I think I got about 18/- shillings. You'd to work yourself up from there till you worked yourself up to £2/18 and you'd all the cleaning to do. You'd over a thousand bobbins to change. You'd your own rollers to clean, the frame ends to clean.

They were all cog wheels and all like that, and then you'd to go underneath and clean all in between the bands that kept the spindles running. That was the work on them.

We were happy like, when we used to go to lunch we'd all to go to one place and we used to have a fight for a seat and if you didn't get a seat you'd to go outside and sit on stones.

It were very rough. But we had good bosses. They were always lively. The bosses 'd help you in any way.

If you did your work right they always used to see as you got on.

It's not like it is today you know.

We used to have to walk. It used to take us half an hour to get to work all through the fields.

We lived in a little village and we used to have to walk to work. One winter it were over six foot high in snow and if I were lucky I'd catch a man wi' a donkey and cart and get a lift.

I worked there till I were eighteen and then I had to come out because I started wi' a bad chest and I went to another firm where we was weaving and winding.

I worked at one place and you know how you do when there's a few together. We'd give a sign that we were going to th' toilets. I never smoked, but I used to go in there for a bit of a giggle.

One time we went in and we were always playing tricks on people.

One day I went in and just as I opened th' door all th' water came over me. One of 'em 'ad got on top o' th' cistern and were splashing th' water out of it.

Two days after I thought I'd do th' same, an' I saw my mate makin' her way to th' toilets, so I got on top o' th' cistern an' when th' door opened I started to splash, splash, splash and when I looked it were th' manager. He pulled me up over it but he couldn't talk to me for laughin'. He told me not to do it again.

I went winding. I'd been there just six months and I'd to work hard.

I'd to work very hard for £3/18 at the beginning. As times went on I worked there for about ten years and then I left. Mi wages were over £5. That was in the thirties.

I stayed there ...I got married from there.

I went to live in Accrington. I went as a bobbin winder at the mill there. I had one hundred ends to run and my wages there , if I worked hard, I could just get over £6. It was good money then. I worked there for seven years and then I went to Blackburn.

I was out of work for a long while ... a good few months and then I got work at Harwood in a mill. Anything were better than nothing.

I'd to go on the train there and then when we got off the train we had up a hill to climb, and I'd started there doffin before I could get on spinning. I did about eight months then I got on spinning.

I was about thirty two then. I was thirty when I had my son. I was ill for about a year afterwards.

That was before I went doffin.

A lady up the street looked after our Michael from six in the morning till six at night.

Well when I came home I'd everything to clean up and see to everything. I'd have to go to bed about nine to be up again at five the next morning and I carried on.

It started going down hill did the mill, so it were closed down. I'd worked there about eighteen months. I went back to the unemployment. We moved out of Blackburn and came to Burnely.

My husband was a weaver all the time. He was never out of work for more than two weeks because he'd travel on a bike as far as he could. He never would be out of work unless he were forced, and that wouldn't be for so long.

When work was bad to get my husband used to cycle all over to different towns to find work. One morning he fell of his bike in th' canal. He were takin' a short cut.

A fellow brought him home wet through. Th' doctor said he should leave th' mill work because he had a bad chest for a long time.

We went over to Smith's Mill and I got work over there. I went cone winding. We went because we never seemed to have any luck. We got an offer and we thought we'd make a change. We were over there five years and then back to Blacks and then they opened another mill near where we worked before, and I went there working, battery filling.

I worked there for two and a half year and that closed down. So I went to another just further down the road, and worked there till I was sixty two.

It was the best place I'd ever worked at. I was cone winding at first and then I did a bit of battery filling.

We were very happy there. You could always go and speak your mind and you always seemed to come off the best way. I were only sorry as I had to give up. My legs went. Going round battery filling I used to do so much and then I had to sit down or I couldn't manage for the day.

I've been all over. There was always something happened
I was never stopped with my work. But there was
some rough places.
But I always thought Williamsons was the best place I
worked at. I was sorry I had to give up, because you
could go and have word with them and they'd sit and
listen to you, and that's what you couldn't get at a lot of
places.
They'd tell you to get out.
I fill me time in now with mi window boxes and
decorating mi house. I go to Bingo wi' mi niece about four
times a week, and we share what we win.
When you've been hard up for a lot of years, it makes you
careful.

Dan and Jane

Dan and Jane are both over seventy (born towards the beginning of the century) and live in a little terrace house facing a factory shed. Their home is like a 'little palace' inside, with a display cabinet full of china in the front room, a tiled fireplace surrounded by brass ornaments and glass animals, and a window sill full of brightly coloured plastic flowers and ferns.

The kitchen/living room is comfortable with two arm chairs, an open fireplace and television. In one corner of the room is the sink and a kitchen cupboard and cooker. They are both sprucely dressed and bright. Dan has a pile of library books on the table by his chair.

Dan. It was grim but happy in the mill in them days. We were there from six in the morning and it was all black stockings, clogs and shawls. You could hear the clattering of clogs down Pear street every morning.
There were a lot of mills round here in our young day. There were about seven just round here.
 Now there's only that across th' road. I started work at twelve in 1914, half-time. I got 3/2d for afternoons and 4/2d for mornings plus Saturday mornings.
Jane. I started when I was fourteen - full time. We used to have to stop in an' clean us looms on Saturday morning.
Th' machine stopped at eleven and we cleaned 'em till twelve.

Dan. I did one year of half time then I started full time when I was thirteen at 7/4d a week. I was tenting. I helped a six loom weaver. Then I gradually got two looms. If your father was a tackler, he would push you in earlier.

My dad was a lorry driver. In my youth, there were only th' pit and the mill. There were nothin' else really. There were only a few other minor jobs. Like a girl could go to one of the sewing firms. But they mainly had to go I' th' mill.

Jane.   Mind you, we both enjoyed it.

Dan.   Oh aye' we enjoyed, every minute of it, I would've worked Saturday afternoon and Sunday, if it'd been possible! There weren't many men enjoyed it like I did.

Jane. They used to fine you i' them days for poor work. You used to have to pay 6p for a seam and sew yourself.

Dan.   Aye an' they charged you a penny a week for hot water for tea.   An' they charged you a half penny for oiling th' looms.

Jane. We even used to have to pay a penny a week each for having th' toilets cleaned.

 Dan. A four loom weaver only got about 25/- a week.

Jane. Mi two sisters went in th' mill an' all and mi brothers went in boat building. Mi father worked on th' canal. Mi mother never worked in th' mill we came from near Wigan...the canal side.

Dan.   I come from Burnley, I never thought of going into haulage like mi' dad. I was always a bit little, and never over-stretched.

I passed to go to th' Grammar School but mi' parents couldn't afford for mi to go. They wanted your wages in them days. It were a case of havin' to get down to work.

I had four looms at sixteen and a half.

Jane. I got four looms when I was eighteen. We met when I went to work at Longbridge mill. I'd be about eighteen or nineteen then.

Dan. We had to court for ten years. We couldn't get any backin'. Work was too insecure then.

Then there were a boom after the 1914 war.

Th' mills were over- capitalized .....everyone were getting' on th' band waggon and then they were closing down, they couldn't show profit. We were out of work.

Jane. When we were out of work, we had to stand in the warehouse before six o'clock. Sometimes, th' weavers were sent out for a week for being late.

Dan. When our Nora were born, I only had two looms. There were no employment - stand back wage. You couldn't sign on th' dole if you had two looms. But still they were happy days.

Jane. I enjoyed mi work.

Dan. When you worked at th' mill, after you'd got your feet in, it were like a big family. It's not like that today.

Jane. Now it's all tip and run today.

Dan. You can't approach people today like you could then. Before, **if** you had a domestic problem, you could go to a work mate or to th' boss. At least you could at Smith's.

Jane. Your hair had to be tied back or plaited then, not like th' young women have their hair today.

Dan. Ah- but there's not th' same revolving belts and main shafts as there used to be on th' old Lancashire looms. I was on four looms for years. I didn't get on to six looms for a long time.

We got married in April 1930. When I went back to work I dropped down to three looms and then to two. We hadn't been doing so bad till then. After that I was off and on.

It picked up a bit after Nora were born in 1932.

Jane. I stayed at home for three and a half years and took washing in. Then mi mum looked after Nora, and I went back to work. When I stopped off and took in washing, I looked after a child who's in the mill now.

After we got married, we lived over a year with my brother.

Then mi dad died and we went to live wi' mi mum for seven years. We came to this house in 1938.

Dan. I had only two or three looms for three years and then there was the six loom strike, and they brought the Manchester police to us.

We were locked out at Smith's and then we all came out on strike. When you had four looms you had to do every job. When you had eight looms you had help. Some of the weavers didn't realize that.

Oh aye, the mounted police came from Manchester.

Jane. There have been some doos in Burnley. At one time there was no work for eleven months, and we were on th' dole.

We were standing at Smith's every morning waiting for any vacancies. If somebody weren't at their looms when the morning whistle went, they put somebody else there.

You had to show your face every morning, and wear clean clogs and a white fent on.

They might pick you or not....you more or less had to 'ave someone to speak for you. You needed someone to plead.

Dan. We used to have to go up Burnley Lane on Sycamore Street to get work. Sycamore street was full of mills from one end to th' other i' them days.

We used to get about 17/- dole. I know it were never above a pound.

Jane. I used to take washing in from two people. I got 5/- from each that was 10/- a week and I took in a child for 5/-- to 7/6.

Dan. But you weren't much i' pocket wi' takin' a child in. You had to give it breakfast, dinner and tea.

Jane. I didn't do that long before I went back to work.

Dan. When it were bad, she signed on the dole an' I worked her looms. I could make more money weaving then 'er. My pickin' sticks used to say, "*Get at it, get at it*". Her's used to say, "*Take you time, take your time.*"

Jane. We were good weavers in them days. We had to be or we were fined for it. We had to pull back any bad work.

Dan. You were a good weaver or you were out of work.
Jane. It was always grade A cloth. Smith's closed down in 1956. We didn't get my redundancy pay. I'd worked there thirty-one years.

If there was any special fabric, I made it. When Smith's closed, I had six silk check looms. I earned more off six check looms than they did off ten looms. I used to weave tartan taffeta, umbrella fabric, tie silk, swimsuit fabric and all sorts. I moved to Scrantons then I went to Hale Dyke. I went on automatic looms, first thirty-six then forty-eight, then sixty-four. I ran sixty-four for a few weeks and then I left.

Dan. I went to Frederick Whites for the last five years to get some new looms going for them. They were just starting with automatics - gabardine suiting it was.

Jane. Changes? Oh aye' there's been a lot of changes. From being easy going' it got all tip and run after. You never had a minute to yourself. For 7 hours you were expected to be doing something every minute.

Dan. With six fancy looms if you wanted a break – which I never did, you could go and have a smoke or a chat, but it's not like that today.

Jane. I finished eleven years go when I was 65.

Dan. I finished at 64. We didn't want our Nora to go th' mill We'd had it grim.

Jane. We didn't want her to have to cope with that. She went in for typing. She's a clerk in the post office now.

Dan. Oh we sacrificed for her. We didn't intend her going in the mill.

Mind you I'd do it over again. I enjoyed every minute of my working life. When I retired at sixty-four, it took me six months to get over it.

Jane. We have a lot of visitors now and we help two old people. I go errands for one old lady up the road, and bring her pension for her and I visit another once every day. She lives just round the corner. I've known 'em both a long time.

Dan. Oh Burnley welfare committee have recommended her for being so good with old people.

Jane. Money? The young ones get money too easy these days. They get 30 pounds to 35 pounds and they can't make ends meet.

We manage on our pension. Even when times were hard, we always used to manage to have us holidays at Southport.

Dan. The young have too much money these days they don't value it. We didn't have it rosy but we managed. We never used to have no holidays with pay. But we went away, - to relatives that is.

Jane. Spare time? Oh I read a lot. I go to the library about twice a week.

Fanny Tillet

Fanny lives in a council flat on the outskirts of town alone but for her two cats and a budgerigar.

Us old weavers don't call these nowadays, weavers. We call 'em machine minders. They depend on helper-weavers. There were no canteens in those days. We took us breakfast and lunch with us. We put us lunch in a basin and warmed it up in an oven.

It was mi brother who learnt me to weave – I was only thirteen when I started. Then he joined up. He was only seventeen. In those days, two-thirds of the people in Burnley depended on weaving.

The tacklers? They used to stop in th' warehouse. Many a time, they used to get behind th' bundles and have a sleep and we had to go lookin' for 'em. Generally, it was a tackler's son that got the job. It was very tight. They wouldn't learn an outsider.

Some people I know used to take in actors as lodgers from the local theatre.

One morning at breakfast, one of them though the cavalry were going down the street, when he heard the clogs of the weavers clattering to work on the cobbles. There used to be a cavalry barracks nearby in those days.

Now they will wear clogs but of a fancier type. These big thick soles – they wouldn't wear them if they'd worn clogs like I did.

Those with babies used to carry them out to nurse, to neighbours or friends, to be looked after while they went to work.

I didn't go to work for two years after I'd had our Jack. When I started work, I had mi' mum living with us.

After th' first World War, we changed to 7 a.m. start. We

thought we were in clover after that. Then we stopped
Saturday morning a few years later.

When I were little, a man was well off if he had 30 to 35 bob a
week.

My dad died when I were six. There were no pensions then.
Otherwise, I don't think mi' mum would have sent us all to
th' mill.

Mi mum was a dressmaker. Mi mum and dad's families
were all farmers. Mi dad was an orphan at twelve so he lost
his chance wi' farming. Mi mum was one of four dressmakers
and was put out to serve her time as a dressmaker.

She said a dressmaker always has to work when the nice
weather is on so she put us to the mill.

Her eldest daughter had only just started working half time
when mi' dad died....Mi' mum had hard times.

She didn't used to get much work in the winter unless there
were a funeral on. In those days, they used to go in black for
th' cat as they say.

Lots of folk used to have their clothes made for 'em. Mi
mum used to send me wi' a note to get buttons and
linings and trimmings at wholesale prices. At night we used
to pick her pins up.

We lived in a four roomed house. We couldn't go in at the
front door for chance somebody were tryin' a frock on.

In February, the new fashion books came in. Then people
used to come for new clothes for Easter and Whitsuntide and
the Fair holidays. After that it was quiet till Christmas, unless
there were funerals or weddings.

We didn't used to have holidays with pay you know, and
some couldn't afford to have holidays without pay.

I've seen men walk to Nelson from Burnley for work. The
married men with families and no pay tried to get work.
Some managed to get maintenance work on the machines in
the factory.

They thought it was grand when holidays with pay came in

I was the youngest child and stayed on in mi' mum's house after I got married. Mi' mum stayed with us till our Jack was six and then she went to live in a cottage on her own.

The young girls now, they won't work hard like they used to do. We used to get a penny in the shilling spending money. We used to tip up all the lot to our parents.

We used to wear a long ribbon with a big plait hanging down the back. Mi mum bought us each a new one in the spring. There used to be band concerts in the park on Sunday afternoons in the better weather, and the lads used to try and pull th' ribbon off us plaits. Then we would have to get new ribbons. We used to put our plaits down under our costumes. A big black ribbon used to cost 2/-. We used to prefer black and it was the dearest. You might wear a pale coloured one to match a suit for a wedding or something.

They usually started wearing shawls at about sixteen.

When you got three looms you used to have a shawl.

Thick black stockings started going out at the end of the first World War and cashmere stockings come in.

A lot wore stockings to match their costumes. I got a brown costume and I wanted brown stockings to match but mi mum would n't buy them.

There used to be a lot of cinemas in Burnley. Children used to pay 1d and 2d. There used to be some good theatres an' all in t' first World War.

Some real companies came. They left London to get out of th' air raids. People used to go straight from th' mill and take their tea and queue. People came from round about as well. They had a good company once for ballet dancing and that.

Nowadays, there's not much theatre or pictures.

The couple next door, they had no family, they used to go to the pictures every night. They said they saved on light and fire.

It's dearer now. A tackler went last week and said with ice-cream it cost him 30/- for him and his family, for a night at the pictures.

After mi mum left us, (to live on her own) our Jack was very sensible. He used to have a key on a long tape round his neck. When he got ten, he used to light th' fire and make th' tea before we came home. Sometimes, he used to come and see me in th' shed and say,

''There all going to th' pictures mam. Can I have some money? I've set th' table ready.'' He was a good lad. He joined this church youth centre and choir, and he used to like a game of football. A lot of these children that get into trouble, their parents don't bother about them.

I knew an old lady about ninety. She started work at ten. Her aunt bought her a white fustian apron.

A lot of the tacklers wore white fustian trousers and aprons. When I started that went out. We had pinafores pleated round at the waist.

We took our top skirt off, and put aprons on. When you were under sixteen, your pinafore had a top.

Nowadays they work in fancy shoes.

Maggie Bates

Maggie Bates is a small frail woman with white curled hair. Her terrace house is on a quiet little cul-de-sac just a few minutes away from the mill where she spent all her working life. The interior of the house is rather dark and drab. She says they are not decorating it as it's due for demolition. Her potted plants, brother's water colors and drawings and her own embroideries are the main decorative features of both her parlor and sitting room.

The parlor is not used much, mainly when her brothers come to stay with her from the United States. Otherwise she uses it for hanging up clothes to air and puts awkward pieces of furniture in it, such as the big wardrobe that she couldn't get upstairs, and the old sideboard that she moved from the living room when she got a new modern one from her brother. Maggie's pride and joy and major hobby is her 'garden,' -wooden flower boxes she has made herself in her back yard. There she grows her geraniums under jam jars. Her one hope is that their new council flat will have a little bit of garden.

When we left school, we'd to go half time - one week at t' morning an' one week at th' afternoons and then we went tenting and when we went tenting, after twelve months when we'd learnt, we got twelve and six a week.
I was thirteen when I started. The looms were very close to. You had to be very careful or you'd bang yourself an' you worked from seven in th' mornings till half past five at night. An' we'd to work Saturday mornings while half past eleven. We had to help this man wi' six looms. We had to do all the jobbing work. Now when we got two looms, I can say we

averaged about 15/- or so for a week.

An' when we'd been on two looms about two years, we got three looms and then it went on until you got four looms. The women were only allowed to run four loom' then.

To tell you the truth, I weren't educated for any other job. We left school when we were twelve and anything like mathematics and spelling or ow't like that – no. I had to use mi hands.

I learnt to weave at Crookshanks, and when I learnt there mam worked there, an' she came to Wardle's and she took me there, and I went tenting when I were thirteen. That would be in 1914. Dad worked i' mill an' all. But he were i' warehouse. He were like a dyer and bleacher an' mi two brothers and mi two sisters they worked i' mill, but mi youngest brother didn't. At first we all worked at Wardle's.

And then the boys went in th' army an' mam worked while the boys were in the army. She 'ad to do and then when the boys came home, she gave over to let the boys go on her looms. An' it were damned hard work at Wardle's – you didn't have time to – it were always on top of you.

You had work for your money.

Mi dad worked i' th' warehouse -dyer and bleacher and when they'd got so many warps ready he were slack.

But to keep him working he'd go on the (mill owner's) farm to help with the horses or go and chop some firewood. You know they kept him going because the' color trade were dying out. They dyed the coloured weft or bleached it, and then if they were slack at Firegate he'd go and give them a hand up there (on the farm). Mi dad had three brothers.

One didn't work in th' mill long. He went to work i' th' corporation. Uncle Will were a painter and decorator and uncle Tom worked i' th' foundry. But there were nothing else hardly for women. They'd all to go i' mill.

Mi' grandma, she worked in th' spinning mill and grand-dad was a weaver. On mi mother's side grand-dad was a cabinet maker. He used to have a shop on Pie Street. He put all these windows in these houses, and they've been in ever since. Mi mother's mother died young after mam were born   I don't think she was a weaver. Well in a way, I think grand ma and grand pa Thomson were a bit more up than the Nailers.

Two of mi' brothers went to America. It was after th' war when jobs were hard to get and they didn't like th' mill after they'd been out of it in the army. An' there weren't a right lot of work.

Our Sam went right after the war, and our Ted's been back every year for three years. And our Sam, he came back last year an' we had him for three months.

Of course, he'll be seventy-nine now and he's going senile. He's worked himself into th' grave. He went out there as a tackler at first and then he got a laundry and our Ted went out there after him.

He played football for a firm, a carpet firm and he came over to play football in England once with his team and we got to see him for two days and since then he's come over pretty regular. He likes the football. You see mi dad were a footballer before ...when he were young. He played football for Blackburn. Then when he'd finished football, he'd go in the mill.

Children? I haven't any. I were too late. Ship 'd gone. I was forty-eight when I got married. Mi sister had one boy. He went to New Zealand and he's done very well for himself. He were a painter and decorator.

He went after he got married. Some relation of hers was going out to New Zealand so they decided to go too. So after they got married, they didn't set a home up, they went. They've been there a long while. He's done very well for himself. He's in government and she's with broadcasting. She makes the clothes for the artists and alterations.

Our Sam's children too are in America, they've done very well for themselves. His oldest son is deputy chief of police. Mary's husband is a car salesman, and Donald, I don't know what you call him, like a maintenance engineer, - very well paid. Both of the boys are in the money there.

One of them came over with his wife and child. Mi brother Sam brought them over for a fortnight. I 'ad four of them, I'm the muggins for the family. I have mi sister living at the other side o' th' town and mi youngest brother lives at Morecambe. When he left school, he went in for a chartered accountant and then he got married before he got his articles, and then he went in an office, and then he went in the war an' he lost his leg. His foot were shot off and then after that he went working for th' Post Office.

In my day, we all went i' th' mill. There were nothing else. We 'ad to go for th' money.

Th' tacklers were always o' th' top of you at Wardles, because if you didn't get your average off the manager brought you up. You had to earn your money.

If they brought you up, they wanted to know why you weren't getting your pieces off; why you were getting so much money. They 'adn't time for them that weren't earning their money. They wanted to know and then if it were th' tackler's fault, he'd get brought up and he'd get chastised.

At Wardle's it was only one of the mills that I know of where we'd no sweeping to do and no weft to fetch and no pieces to pull off.

We took the rollers out. But they stopped so much out of our wages for that. We were supposed to do nothing else only weave.

After th' first world war, they went o' th' six loom system but Wardles' didn't go on straight away, because they'd such a lot of dobbies and coloured warps and then eventually they went on t' six looms, either five or six looms.

The men were there — but they gradually faded out, but not for a long time. It were a long while after the men started floating out.

It were a mixed factory and it were hard work when you had six looms, because in Wardle's you had a lot of fancy cloth.

The people at work (while I was at the mill) they were nearly all the same lot. Once you got in the mill, you had to stick it.

An' it weren't often they got a new weaver because when they went on six looms instead of four, they got rid of them, they didn't want.

They kept them on that worked to run the six looms, an' then after, when things got eased off, an' men had gone into the engineering and what-not places, they couldn't get weavers, anybody could get work then.

At one time when things were bad, we went on a three day week. We got three days unemployment.

When the coal strike were on, we had to stop.

Mi dad were going round buying coal what they'd picked up, to let us run three days.

Mi dad were like th' bosses' handy man, kind of thing. He'd tell him to do something and he'd do it.

An' when they were digging for coal, he used to go out there and buy what they'd got to keep us going so long, and then it faded out and we had to stop.

Oh yes, we went on short time two or three times, but Fred Wardle weren't in th' Masters Federation and when there were a strike on, Wardles' run, because with him not being in the Federation he was on his own. We were lucky that way.

There hasn't been a lot of cotton strikes. Last time there was one we were all working. They couldn't interfere with us.

It were a damned hard life for a woman, an' they all had their homes to run, and there were no vacuum cleaners anything like that. I don't know many people workin' at th' mill now, you'd be surprised. I still know Long John and Old Bob Simpson. I went to see Clough last week i' th' warehouse and they were all strangers.

I'd like to go in the shed and see them new looms them self threaders. I think th' manager 'd let me in to see 'em.

But it's not same now as it were before. Like the tackler as I worked under he had four looms a' th' side o' me and his wife. We've always worked together. He were alright, I could always say anything I wanted to him, an' his wife was a supervisor, I could talk to her an' all. I could say what I like there were no dictatorship.

Dick Ashford was th' manager and then he retired. Nobody liked him. He were a dictator.

An' then you see, Fred Wardle, he weren't i' th' Federation. Well we had sweepers and that. Well all th' others got Saturday morning off. All the other mills but we had to work Saturday morning. He could please himself.

Then Dick Ashford were taken ill and Carstyles took over. I can't say as I worked under a nicer chap,- a fellow that you could talk to and he'd listen to you.

Mind you I've known him ever since he went i' th' mill. Him an' mi' dad 'ad a lot i' common wi' th' farm and Fred Wardles. I could talk to Carstyles. He were a gentleman. But I can't say that for Dick Ashford. He were a boy what 'ad been brought up from Blacklea road school, an' he'd got up, and up, and up, an' "*I'm it.*"

He knew everything. Nobody liked him. He liked to show 'is authority. Well I thought so.

It weren't like that 'ed been brought up i' th' mill. He'd been brought up i' th' office and he were different. He came from school to th' grammar school and then to th' office. He'd never worked i' th' mill.

His son Bob, he were being brought up i' th' mill. But he were a grand lad. He started out a' th' bottom did Bob, but he didn't stick it. He went out to Rhodesia to his uncle, but he couldn't stick it there either. He came back and then I lost th' end of him. He were a grand lad.

I wish I were younger, because now they are earning good money a lot of different money to what we earned, when we were there before.

I liked the automatics – admitted, I were getting old an' it was hard work. But I kept up, an' I'd go back tomorrow if I were able, because it's the only work I know. A n ' it's one that you like, an' when you like a job you keep it.

 An' I will say this – to me it was more comfortable on the Lancashire looms. You knew what to do. You knew who were who. A n ' they weren't coming a' th' back of you.

If a tackler or anybody come in and saw you talking, they'd stand and watch you, whereas when you were on th' automatics, you knew you had your work to do and you did it.

Yes I'd go back tomorrow if I were younger. Because, I mean to say everybody don't dislike weaving. Somebody has it to do. I didn't dislike it.

The only trouble wi' me was getting up i' th' morning.

At first, we had a knocker up, when we were all at home. We 'ad to do.

An' if we were five minutes late we all got sent home, - the lot of us. Yes we got sent home for the day.

When we were all working at home, we had to manage as best we could. We had to pull us whack, and we all had so much work to do. I 'ad to do out the back, and the lavatory, an' clean mi' dad's shoes, an' then o' th' Friday mi sister 'ad to do th' parlor. We'd all to whack in.

I' them days we'd black leaded ranges, big ranges.

Well I'd to clean all the ashes out, an' then mother used to black lead, an' the knives and forks were all steel. I had all them to clean wi' emery paper. In a way, nowadays they don't know they're born 'cos there were no vacuum cleaners then. Oh we all 'ad us work to do.

 An' then when we were at home, at breakfast time, we had to come home, an' mi mother 'd be bakin', an' after we'd had us breakfast, we had to kneed 15lbs of flour. Wednesday was baking day, we knew what we were coming home to - potato pie and rice puddin'. Ah' we had to pull us whak.

We didn't get chance to go out. We got very little spending money.

We went to' th' penny pictures. Rialto was two pence. And then we started going dancing. We used to go dancing at the Conservative Club, and that were six pence to go in, an' then we got older they opened the Empress.

We used to go down th' Empress. I think that were one and six to go in, and then it went to half a crown. Mi' mother taught us to dance. I've danced all mi' life. I dance yet. We go every weekend Saturday and Sunday.

I learnt Willy Clough to dance. We go to Burnside Conservative Club. We're members we don't have to pay. An' we dance while quarter past eleven, an' on Sunday it's dancing and Bingo.

I don't go out i' the week time. I can't afford it, when we're on th' pension' We don't go out thro' the week.

Mi husband, Albert, goes out on his own on Friday night, while I wash mi' nut (hair). Saturday an' Sunday we're up a' th' club.

But when our Joe were here we were up and down at football matches. We went all over. I'm interested in football and cricket. Well I'm interested in all sports.

I never played any games mi self. We hadn't time. Th' only sports we had, was Church walking day, - bun and coffee day. And we used to have races. But mi brothers were all keen swimmers. They can all swim but me, and our Joe's always been interested in football. He had a trial for Blackburn. He played for them a bit. He took after dad.

When I got three looms, I got half a crown. I had to buy my own stockings an' save up for me holidays to Blackpool. It were all we could afford. I went wi' mi' friends for a week. We only got a week's holiday wi' no pay.

They used to run a bank a' th' factory – savings bank. I don't know whether they do now. But they always have done. You paid your bank money in when you drew your wages.

Sometimes you could put 6d in sometimes you'd put a shilling in, they never bothered what you put in, but you couldn't draw it while Blackburn Fair, unless it were a emergency. You were dam lucky if you had three pound in. In them days, you used to have apartments. You just paid for your sleeping and kept yourselves in them days.

Well, everybody hasn't had a life like me. I mean to say, Mam had to work hard all her life, an' we hadn't a father as saved anything, because he liked his beer, an' he was off.

I wouldn't like a life like my mother had. I always said I'd never 'ave a life like my mother 'ad. I must say this Alfred (mi husband) 's been good to me wi' money matters.

I didn't get married till late, because I had to stop at home wi' mi mum an' dad. In a way I were th' main stay at home, to keep things going. An' I promised mi mam I'd never leave her, and then before she died, I promised her I wouldn't leave mi dad.

I've had some good times. They haven't kept me down. You see there were four of us went out together. We were all of the same mind. We were out for a good time. We'd go to th' dance and come home. We didn't bother, we didn't. We had a good time. I've lost touch wi' them now. I think two are dead.

Mi' school friend. I write to her yet. We were at school together. We both tented at work together an' we'd three looms together, and then when she'd learnt, her father bought a lodging house at Barrow in Furness and she went up there, and I used to go up there for mi holidays. That's why I never bothered. They sold it just when the war started.

They hadn't been out of it for a month when it were gutted. An' she lives down Bournmoth now, an' we correspond at Christmas. We always send one another a card. With her being down there, we don't see one another now.

An' we've had some happy times at work, especially at Christmas. Some of the women 'od go round kissing all the men.

An' all that went into the' Sports Club, we all joined at it an' we used to have two dinners, one in the summer time. We'd go to Swanmere Hotel and we'd have bowling handicaps and then we'd have dinner, and after that we had the room upstairs, and then we had a big singsong and a party. And then the sports club used to take us.

We'd go the Blackpool. But one of the best does that we had was when the Exhibition opened at London and the firm booked coaches, and there were Firegate Mill and Wardles' and we had coaches to take us to the station. We went on t' train to London an' when we got to London we had breakfast, and then we'd coaches.

We went to th' Houses of Parliament, St. Paul's, Westminster Abbey and the Tower of London.

Then we went back for dinner, and then after dinner, t' coach took us t' th' underground an' we had th' rest o' th' afternoon and night to ourselves. But we had to be at the train by half past ten, an' we went to Battersea Park. We had a good day there, an' we went a sail up the Thames, an' we came back, an' we took the train home. They gave us a good day out. Mine were free, but I'd to pay for Alfred. I think it were 35/- for Alfred an' we had a right good day out.

An' then the following year, we had another affair, an' we had coaches to take us to Blackpool. An' then we had dinner and tea at the Casino. An' then we could please ourselves. We had to catch the coach back at eleven o'clock.

That was the two trips that we got out of Wardles'. An' Christmas we used to have parties in the mill.

We'd all pay three pence a week for so long, and we'd have meat pies an' cream cakes, and we'd join at a bottle of wine an' we all sat down at half past three, an' we had our party in the mill. An' the management didn't say anything.

Them as wanted to work could work. Them as wanted the party had to stop their looms, sit down and have their party. They didn't interfere, but you'd had your party for that half an hour you'd to get back to work.

I were very late getting married (at 48).

I were weaving six looms then. When we got married, Alfred worked at th' foundry and then his eyes got burnt an' I made 'im come out. He got work wi' th' street lighting o' th' corporation. Well he's retired now.

After mam and dad died, I were on mi own about three years and then I met Alfred.

I were born across th' road and then we moved to th' house next door because there were a parlor, an' mi' mam could make it into a bedroom for me an' our Jane.

And then when the boys went in the army- of course they were men when they came back - we moved on Belmont Road an' grand-dad were wi' us and then after the war, our Sam went to America.

Our Tom got married and he went to America and mi' sister got married. And our Len were getting married and mi' mam says, ''*Well, these 'ouses belonged to me aunt*'', and she said ''*There's a house comin' to let down yonder Maggie,*'' she says and we moved in. It were this one and we've lived in it since th' Second World War.

When me mam were poorly before she died, I'd been off me work. They were on short time at th' mill. An' Fred Wardle put me o' th' dole so that I could stay at home an' look after mi' mother an' so as I'd have a bit o' money comin' in. I were at home three years with her and then after dad were up at the Wardle's Farm an' he had an accident up there.

He fell and fractured his skull an' Carstyles brought him home in his car. On the Sunday, Bob Harrison from work came an' he says, "*Get the doctor Maggie I don't like the look of your dad*", An' the doctor came an' he ordered him to hospital an' he just lived a week, -fractured skull. We never knew what happened to him.

An' I never got a penny, 'cos we don't know how it happened. Fred Wardle said he'd look after me, an' as soon as dad died, I went back to work. All we got was a big wreath.

 I've nothing to thank them for. I've had me livin' out of 'em that's all. You haven't to reproach yourself for what's gone. An' it's no good livin' with the past.

When th' factory were on short time at work they always kept me dad on, because if he weren't there 'e were on th' farm, Fred Wardle's Farm an' 'e 'd look after the horses an' if they wanted and jobs, he had to do it. He'd to go and chop firewood for old Billy Wardle, Old Smiler.

An' when they killed any hens or chickens or what not, I 'ad 'em to pluck in mi' kitchen, an' he'd take 'em up after.

But still they kept him workin' an' you'd to be thankful for that in them days.

Fred Wardles, he used to walk into th' shed twice a week with his billycock on and his hands behind his back. I don't know whether he used to see anything. In the mornings, he used to ride up the park on one of his horses before he went to th' mill.

I remember I once went to see mi dad i' th' stables, when he were grooming th' horses, and I saw a lot of old patched underwear hanging on a string -like a patchwork quilt they were- mended with darns and patches.

I asked me dad whose they were. I could hardly believe him when he said they were Frank Wardle's. He was mean. He never spent a penny. He used to be quite fond of his horses though and mi' dad used to take them to races. An' he must have been worth thousands.

I'm the oldest living round here. The old lady across there, she came in that house when we flit and the lady next door, Sally, she used to live across th' canal. She were brought up down here. Lily were brought up down here too. I've known her all mi' life, and Mr. Ronald across there. I've known him ever since he came from Blackburn as a baby. His father came to work the skin yard from Kendal an' they were on th' front street.

They were lookin' for a house and mi' mother brought 'em in and she let 'em sleep in the living room and that. An' she got 'em a house. I've known him all mi' life, but I never bother with them.

I acknowledge them, I speak to them, but I go in Sally's next door. She's poorly, an' her niece has taken her away.

I've been up this morning to Each Street. Her hearing aid broke, so I took it to the hospital on Monday, an' it came back here, an' I took it to her house this morning.

Oh, we all help one another. We won't see anybody stuck. The lady across the road, May White, I've known her all mi life. She set th' house o' fire th' other day. "*Where's help? Come on Maggie.*" And she were laid o' th' floor. She's poorly. We got her right. We all help one another. Well we are like that. The others I don't know. I don't know any of them.

All these houses have to come down. It's not classed as a slum house, but it's in that area that has to come down with the rest. I mean to say, when you've to start paying rent.

We've had us money's worth out of it. But we've known for three years it were coming down. That's why I won't have no decorating done.

We'll get compensation. We'll get the value of the house. I've seen to that. Up that street and on that side, they won't get it.

Across there they'll all be pulled down as slum property.

You known I'm not bothered in a way.

They might put me somewhere where there'll be a little garden out back. I'd like a little garden.

We'll be like everybody else we'll get through. God's been good, we'll get through.

One thing I will say since we've been married, we've had some good holidays. We've been to America for a month. My brothers paid for us and then we always had a fortnight away. We always went somewhere every year.

We've been to the Isle of White twice, Hastings, Clacton, Brighton, Eastbourne, Newquay, Ilfracombe, Great Yarmouth,

Scarborough, Llandudno. And last year we went to Cleveleys for a fortnight at September.

Oh we've had some good holidays since we got married. We've spent money on ourselves.

It was five years ago we went to America to see my brothers. The supermarkets are lovely. They are four times as big as Blackburn market hall. Oh they're gorgeous.

But you have to have a car. If you just want a packet of cigarettes you have to go as far as from here to Manchester. I wouldn't like to live there. It's too fast.

We stayed at a lot of different places. We went to Washington. We went up the White mountains. We went to a place where they breed trout and what not. Oh! it were interesting. But I would have liked to go to Niagara Falls, but our Sam were too old to take us.

This is a photograph of my niece's wedding over there. The men had their dickie bows to match the bridesmaids, and the flowers matched the bridesmaids' dresses.

They have a different do altogether from what we have here. They asked us to go again at Christmas but we didn't because our Sam was ill, but I think we might go in the summer. We haven't booked us holidays yet. We're waiting to see.

Last year when our Sam came for two months, we had a car at the front door. We hired a car to take us out. Our Sam wanted to go all round Blackburn to different places, where he went when he were a boy. He were living in the past.

Last year, we couldn't afford a proper holiday, I went through ninety, no seventy pounds for food and entertainment. I'm th' mug. We could only go to Cleveleys at September.

We're on the pension you know. It's not food (that costs a lot) it's cigarettes. They're my only pleasure and I'm having it.

Our Sam's painted that picture on the wall, he's painted all me paintings, an' I have 'em up.

It's very expensive in America. We went to a motel with our Sam and it cost nearly ten pounds each for me and Albert for a motel.

But it didn't cost me nothing. They paid.

And you only sleep there. And then you have to pay for your own breakfast. It's all pay out there.

It's very expensive, and by God don't they like their steaks! I said to Sam are you going to eat all that? Three of us could have sat down to it. I've never seen anybody scoff steaks like they do.

Well our Tom when he came, he took us to Leeds to the football match. He paid all us expenses. When we came back we went in a café i' th' town. We had our tea there, I didn't have to come home and cook. Well it were a holiday worth having were that.

He were over this February and he managed to go to th' football match cup tie. An' it were that day that new stand opened, an' we got seats in the stand, an' he took his ticket back with him as a souvenir. He's football mad.

Well, he'll come again next year. But he can come any time it's a pleasure having him. He's a grand lad.

He brought Albert that shirt hanging up there for £5/10. He wouldn't have had it if I'd it to pay for. I said he wants a striped shirt.

He took me in town he said, "*How do you like that?*"

I said, "*It's too much money*", I said, "*£5/10 is a lot of money*". He said "*Well, I'm buying it*". "*You aren't*".

I thought, "*Well, if you want to spend* it". An' I said, "*I wanted a new dress*", so he said,

"*Come on, we'll go down an' look for a new dress for you*".

I said "*You're telling a lie. You won't 'cos you've spent enough*". Anyway, he's a right grand lad.

Our Sam 'ud cut a current i' two – different lad altogether.

Our Sam and our Annie are alike. It's all save…me an' our Tom we don't care if we're enjoying ourselves. That's all as matters. We have a bit put away. We're not o' th' beam ends, but we're living off us pensions now. We'll have to draw out if we want to go on holiday. We're not bothered about that.

 If they send for us to America again, we'll go.

 Stop at home as a house wife you say? I never wanted to. I couldn't afford to. Not if you wanted to live and have your holidays. I retired when I was sixty. They bought me a gold watch an' I got a check, but I've never told anybody what I got. The manager told me not to tell anybody. Alfred doesn't know what I got. I worked there all them years – forty-six- and I got 25 pounds! I think there was only me and another girl as got one.

How do you like my geraniums? They're my hobby they are. I made those wooden pots myself. Alfred he just likes to watch the Teli'.

I embroidered all these mats an' all. I make some every year for Christmas presents for friends.

Clarence and Ethel

Clarence and Ethel live in a neat little terrace house ten minutes away from the mill where they spent most of their lives. They are both retired and in their late sixties now, though Clarence works a few hours a week at the mill just to get out of the house and supplement the pension. They have no fancy furniture or luxury items in their house. The walls are decorated with their son's line drawings and paintings and pictures of their daughter's graduation and wedding. They are both thin and white haired, with pale, sensitive faces and expressive gestures.

Clarence . I don't like bullying, but discipline is another thing. It's needed in any workshop. In the old days the bosses were always right. You weren't. But these days the manager cooperates and talks to you.
 You used to be frightened of going home to tell your mother about stoppages. The wife was fined 2/- one week for using wrong weft.
We used to get a penny i' th' shilling spending money out of our wages. Some used to take th' coppers off their wage and keep it, they called it coppin.
Ethel . We used to get our wages in a tin from th' tackler.
Th' wage tin had a number on it. If the wage was poor, th' tackler might throw it at you and you wouldn't dare look 'im in the face. When we were older I used to wonder why we both worked at th' same place, because it always used to mean that we suffered from the same short time.
We both used to come home together for breakfast - tea and bread.

We used to come home for lunch together as well.

Sometimes I went for chips and I'd only just bought them by the time we had to go back to work.

Clarence . When the children were going to school, mi mother used to come to th' house in the morning till we came back at breakfast time and the children went to school.

Ethel . I stayed at home with our Ronny till he was four years old. He were a difficult child. Then I went half days. At one time, I used to alternate with his sister.

Clarence . In the nineteen thirties, there were five years when we never went out. We couldn't afford it. The weavers' strike happened before we got married.

When we were courting, I walked with a banner, some marched from Manchester and the police were waiting for them at Rosefield. I got chased wi' police.

Mi' dad? He were down th' mine like his dad, my mother were a weaver and hers were a winder.

I think on th' whole things have changed for th' better in th' mill.

I started work at the age of twelve, I went half time learning, and then I went half time tenting for a six loom weaver. I earned five and six then.

Then when I was thirteen, I started going to work full-time.

Ethel . I started full time when I was fourteen with a six loom weaver tenting at 11/- We worked five and a half days a week then 6 a.m. to 5.30 p.m.

We worked Saturday morning till 12. We worked in very bad conditions. The looms were all cramped up in a small space. Some looms only had eight or nine inches between them. Anybody fat couldn't get past. In those days the looms had straps.

At that time - the first world war and after - women with babies had them minded out by their mothers and relatives.

Some went out to feed them at breakfast time. Some had them brought to work to be fed.

Sometimes, they were fed at the side of the loom.

Clarence. They fought - the unions to get shorter hours. The weavers with six looms were mainly men. They were often very awkward and strict they used to kick your ankles with their clogs.

Th' wife let an end run down and th' weaver she were working for said she needed glasses.

Ethel. Mi' husband dare hardly go to the toilets for a smoke in those days.

The tacklers were very mean. The tacklers liked to pick their own weavers. With a good set of weavers their own wages would be better. They'd bully you and get rid of you. If you had a nice face and they liked you, you were alright

Clarence. The mill was full of favoritism from managers to tacklers and tackler to weavers, till a new manager came and was very strict and fair.

Before that time, the best warps (easiest to weave) were given to the managers' and tacklers' favorites.

Ethel. We got married in 1933. Clarence had six looms then and I had four.

On the six looms, he earned 30/-. When any looms were stopped because of lack of warps you got no pay.

At one time, he only had four looms going. We had to pay our lodging money out of it, and then bring up a child.

Clarence. I often used to go to work off bread and treacle. At one time, I couldn't even afford to put new irons on mi' clogs when they needed it.

Ethel . Our wages were deducted for faults. For a float we had to pay a 6d fine. If you got t h ' wrong sort of weft in you were only paid for half a piece. You were fined for faulty cloth. I earned 25/- after we were married.

Mi husband's sister looked after our Mary.

I paid her 10/- a week nursing money. I used to go to work heart broken at leaving the child from Sunday night until the following Saturday morning.

His sister's husband had a good job and they were quite well off and she had no children of her own, so she asked if she could look after the child while I went to work.

She was so well looked after by her aunty- she was like a little princess. She used to cry when she came home to see us and our conditions.

So I worked for 15/- after the nursing money was deducted. I'd been on the dole for some months after the baby was born, but when the dole stopped and the baby was ten months old, I went back to work.

Clarence. At one time in the depression in the thirties after Mary was born, I was on the dole getting 18/----15/- for a couple and 3/- for the child. When you went for your dole money, you were treated like dirt.

It was a most shameful thing. We had to live off 24/- for quite some time. At different times , we were on one two or three looms.

Ethel . I used to wait in the warehouse for work and the book-keeper used to signal to one of those in the queue waiting to go in and do one day's work.

Clarence. It was a most degrading period for anyone — most degrading. The week's summer holiday- you dreaded it coming round as there was no money. One day, we walked up th' park. We hadn't a halfpenny between us.

Ethel. I borrowed 10/- from mi mother and paid it back at a shilling a week. On the Friday we broke up for the Fair Holidays we used to have a treat of strawberries and cream. I t was cheap then

Clarence . I even used to go up and down picking up cigarette ends to smoke. I couldn't afford any, and nobody had one to give

Ethel . In those days, you would clean flues for a shilling. The funny thing was that we were very happy. We would set off with the pram on a Saturday afternoon in the country, and take the pram up Bluebell Wood. We used to carry it over styles and hedges.

The reason why we were happy was that we were in love with each other.

I used to go over to tell mi mother mi husband loved me. On Sundays I would do my washing and bake my own bread.

We were paid on Wednesday. On Friday, I might borrow 2/6 off mi' sister in-law to buy some stewing steak for the weekend. But I never got really into debt.

I only borrowed a bit from mi' mother and Clarence's sister, and paid them back. I remember one day in 1938, mi' husband came in.

He'd been out of work then for six months. He came in with a hat band and badge -uniform for our Mary to go to the private convent school. She was four then. He said he had enrolled her there. He didn't want her to be like us.

Clarence. I remember how James Whittaker used to come round the mill every morning with his hat on and walking stick, smelling of whisky.

He used to look at the weavers like dirt. His uncle used to come round as well and point at bits of waste cotton we should pick up.

Ethel . I remember one time when I only had one loom and I was reading as well.

The manager came round and called us to the office and said, "*Next time I catch you reading, you're sacked.*"

The weavers round me didn't dare to shout at me and tell me he was coming, they were afraid of him.

Clarence. I had to wait mi turn to be an over looker. I would be thirty -two or three when I got chance to learn. As an apprentice you had all your own warps to carry. Some were 200 or 240lbs. I had to carry them on my shoulder. Some weeks I carried about thirty. Because the flanges (of the looms) were so near you had to lift your leg up to get over them, bent double under the weight of the warp.

In t h' nineteen fifties, there were more periods of short time. For eighteen months, I worked three days and then four days for twelve months.

Today, it's a piece of cake. If I were twenty or thirty years younger I would go into it. I was fifty-two when I learnt on the automatic looms.

Ethel. It becomes part of you after a while.

Clarence. You always like it.

Ethel. You feel as **if** you have no brains at all part of the time. I sometimes think, why didn't we have the courage to do something else.

But then when we were young, there was only one thing-cotton. Whittaker's was the hardest mill there were. There used to be about six mills just round us. Whittaker's had a lot of bad twist (warp).

Clarence. Tacklers in them days were numb. Some were big ignorant bullies.

It was a closed shop with th' union. Your father had to be in or a great friend to speak for you.

I was lucky I told a lie to get in, I said I had two tacklers to teach me when I hadn't. Some got in before me as started when I did.

I'd been waiting a long time.

Ethel. I had four looms under a bad tackler. When our Mary was little, we had no house of our own, we lived with an old aunt.

On Saturday afternoons at 2 o' clock after we'd finished work, we used to wheel the pram to Bluebell Wood with sandwiches and get a jug of tea.

We were very happy, then Saturday afternoons were like a holiday.

On Saturday nights, I used to exchange a f i l m book at the second hand shop on the corner and get two nutty rings for a penny.

We had a relay speaker for 1/- a week and we used to dance to th' music. On Monday morning, we used to be hungry.

Clarence. But was everybody happy at the time like we were?

Ethel . I think we were happy because we were in love. A lot of people, especially those with a lot of children were suffering.

After they'd paid their rent, they'd have nothing left. They'd live off bread and jam.

I had a routine for th' housework. Monday was wash day, Tuesday I did the ironing, Wednesday was cleaning downstairs, and Thursday cleaning upstairs. Thursday night was the night of the week that we fell out with each other as we were so tired by then.

But we were very lucky. Other couples when th' man and wife were both working in the cotton trade, they used to set off at 5.30 a.m. leaving the children with a nurse on the way.

They'd push them there in the pram, and call for them at night. Some weren't looked after well but our Mary was.

Clarence . It weren't human living. You were lucky if you got chips for your supper without fish. Now I've worked in th' cotton trade for 56 years. At the moment, I work part-time and would like to give it up. I'm nearly seventy

But I'd have nothing to do at home only sit by the fire. I worked for Whittaker's all that time and when I left they gave me 26 pounds. I don't want to say much about that, it worked out at less than a farthin' a day!

Ethel. The firm doesn't appreciate its workers. It's changed today though. It's just the reverse now. There is a terrible lot of easy money.

We used to go to work in clogs and white fents and shawls and no make up.

I remember one Christmas I put on some lipstick and everybody pointed at me. Now you would think they were going to a party when you see them going to work

Clarence. In our day, they didn't think about money, they thought about work.

I remember the humiliation of signing on the dole. They might find you a job and stop your unemployment pay. You felt like a little animal.

Ethel. I told mi mother I wouldn't go in the mill. I was the oldest of five children, so mi mother took mi down town to a florist shop,

I told her I would love to work with flowers. But they said I would only get 47/- a week, so mi mother wouldn't let me go there.

She said I must go in the mill. First of all, I started learning for nothing. Then I got one loom then 2 for three years, then 3 for two to three years. I was nineteen before I got four looms.

They would only give you four looms if there was no married woman waiting.

Tenting? It's not done now. They stopped saying that 35 to 40 years ago.

When we were working, we used to lip read and use hand actions to be understood in the shed.

With the old looms, you couldn't hear a thing with the shuttles banging and the noise.

For example, they used their fingers to show the time. They pretended to put a collar round their neck to show it was time to go home.

Clarence. On Saturday morning, we swept th' looms. We got three pence from the weaver. We got sweepers later and paid them 6d of our wages.

All my family were in the mill. I had four sisters at Phillips, all at the same-time.

Ethel . I had one sister went in th' mill.

Mi' sister in fact miself and the others as well, we were all good at school, but we couldn't go to the High School. Mi mother said no for the first and second. Mi sister never did anything in the mill.

She always earned low wages and was never interested in the work.

Clarence . One of mi' sisters used to come home jiggered. She was married and used to call at mi mothers' like that every night, and then she used to go to bed when she got home every day. And then she'd get up at about ten o' clock for a cup of tea and a bite. She worked under a very nasty tackler.

Ethel. We used to spend so much time taking twist up and we used to put all us own traps in.

Clarence. If them type of workers were in the mill now they'd have 90% efficiency and be there on three shifts now. Every woman was a worker.

Ethel. I did three years on automatic looms. When our Mary was doing her 'O' Levels, we worked on opposite shifts so that there'd always be somebody at home.

Later on, I went to work on broad sheet looms at a village mill, (I'd got too tense working on the automatic looms). The weavers in that village thought th' bosses were marvelous. They had to carry their own weft down some steps and they cleaned their looms with paraffin till they shone. I gave up four years ago.

Those village women weavers daren't lift their heads up when the tacklers went past.

They had to oil their own looms and pull their own pieces off. There were no clocks o' the looms and they used to get three pieces off a loom in a week.

The place was full of mice at night and one day one came out from under a loom covered in fluff and all the weavers except me were frightened.

Clarence . It's still hard work for a tackler at Phillips' if they do their work properly. But they don't.

It's a change from 32/- in my young day to 32 pounds. They don't do their work as they should. There's a lot of dodging. Phillips' 'as always had a hard name. It still sticks today. It was a good place to work for.

Since Jack, the present manager started, you get every penny you earn. Jack always sees to that. You didn't get it in the old days. If the manager had workers to co-operate with him, the place would run well now. The place used to run like a machine when Jack started at first.

There were still some of the old Lancashire loom weavers and tacklers.

It's a new lot of workers now.

The present day workers strike for this and that. I don't know how they sold the cloth in them days.

The twist (warp) wasn't good especially for the poplins.

One tackler was as thick as a brush. A shuttle flew out and hit a clock on the wall. The tackler got a nail and moved the clock! Shuttles used to keep flying out. One young woman lost her eye.

Ethel. I remember seeing her walking out with her face bleeding.

Clarence. The bosses daren't used to tell the tacklers off. They were frightened of them. They had a strong union.

I've seen so many changes. Many young tacklers have come, learnt, and left. They were in their twenties and thirties. They were fools. They didn't want discipline.

Ethel. His mother used to stay at home to look after his sisters' five little boys while they worked in the mill. She used to run 'em round wi' th' long brush.

In them days, they had brass fenders and leaded fireplaces and no carpets, and stone 'earths, but they used to get the housework done. They seemed to be stronger in them days.

Clarence. Good luck to th' youth of today. I was always happy in cotton and I still like my part-time job. I used to enjoy tacklin'. It should be put down on records t h a t fifty years back women of Lancashire in the cotton industry were the backbone of the industry.

Ethel. They were the backbone of their husbands too!

Clarence . I give them credit for it.

Ethel. No woman liked to go to work and leave her baby. Some left two or three. Then they came home and did their housework. Granted, some were rough and ready.

It was all bed and work fifty years back.

Girls on a Saturday afternoon used to have their curlers up to go dancing on Saturday Night.

In those days, there were six or seven dance halls and a skating rink. It cost 3d or 6d to go dancing. I remember borrowing aunties' shoes and stockings to go out in, and I once lost her umbrella.

Clarence. People used to be a lot more neighbourly than they are now.

Ethel. I remember when mi mother's fifth baby were born. I was eleven. Mi mother had a bed downstairs and I was looking after her.

And every day neighbours used to bring meals in, a bit of steamed fish and that.

Clarence. When the woman next door used to bake anything, she would always go round with an oven bottom or a sad cake.

Ethel. There's a big difference between our generations and the young today.

Clarence. Our son 'as been out of work now for two months. He doesn't like working regular hours. He works for a few months and then stops.

He lives at home and pays board to his mother and then he stops off work for some months doing what he wants.

Ethel. He likes writing and painting. That's one of his on the wall there in the front room

Clarence. I don't understand him at all, the way he carries on.

We're proud of our Mary she got her B.A with honors at Liverpool. Then she did a secretarial course.

She hadn't worked long when she got married to an American teacher and went to live in America.

Ethel . She only does a bit of part time secretarial work now. She has three children.

Clarence. I'm very disappointed that she didn't go on in her job. She was doing well. She always used to want to be a journalist.

We visited her two years ago. They have quite a nice home, and all sorts of people visiting them.

But I was glad to get back, I'd soon had enough.

Ethel. I would never like to have to live my life again. I can never think why we didn't have sense to leave the mill. I would never go back.

Clarence. Working in the mill is like a virus. It catches you, I would go back.

Adam Hope

Adam Hope is a retired over looker, whose ill-health prevents him from taking on the part-time job he would like. He blames his sicknesses, of hernia, bronchitis and subsequent heart trouble, on his years of overwork in the mill, when he did hours of unpaid overtime and spent many nights and weekends doing union work.

My earliest recollections are of going to the village Church of England School.
There were two classrooms - the big room and the little. I remember going to school in petticoats.
The teacher didn't understand how I could do sums but I couldn't do multiplication tables. It didn't take me long to learn.
My only sister left school half time at 12. My elder brother worked at the local mill.
The mill owners, the Beadales owned every house in the village.
Somebody in each house had to work in the mill.
Two thirds of the houses were back to back and they all had big black fireplaces and stone floors with the sink in the kitchen- no bath. Some were only one room up and one room down. There was three rows of houses.
Half of the top row was lobbied houses and then the next half there was one up and one down at the front and one up and one down at the back.
And the sink was in the house. There was just one bedroom. You just had one sink under the window downstairs for washing up, doing laundry and washing yourself.
There was an oven and a hot water boiler and you'd to fill the boiler with water and shove coal underneath it, an' then we

had a big fire and you could let the bars down if you wanted
to put pans on.

My dad stopped work down the pit when he were about
forty-five. He had bad bronchitis and asthma, so my mother
worked at the mill and kept five of us. And then my elder
brother and sister left school and went to the mill- half
time at first.

 My mother worked from six in the morning till five thirty at
the night and Saturday mornings as well, a full sixty hour' a
week. We hadn't enough money, so she took in washing
two or three nights a week for people who were a bit better
off than us. We used to have a coconut matting down for a
carpet. It was put down at Saturday lunch time and rolled
up again on Sunday night.

Although we were poor we always had enough to eat. Ma
saw to it that, even if it meant her being a bit short. Mind
you I remember three of us used to share at an egg.

We had Sunday clothes which we used only for Sunday
school twice and church once and when we came home, we had
to change.

We were never allowed to play any games on Sunday.

I well remember ma polishing the old black fireplace and she
would use some whitening for below the bars, which always
threw a lovely clean white gleam from them.

She also acted as midwife and laid people out who had passed on
for eighty percent of the village. She never got any pay for doing
that.

Father Christmas never entered our house but we could not
possibly have loved each other more.

When my brother and sister started working at the mill, the
financial situation got a bit better. But wages were very low and
short-time work was always in the offing.

My brother started helping on a farm a couple of years before he
went in the mill. When he went in the mill, I took his place. I
was about nine then.

I was up at six every morning and went to help to milk the cows and feed the hens. Then at about quarter to eight we went round the village delivering milk.

We generally finished about ten to nine and then I had to go to school. When school finished at four, I went straight to the farm again to milk and feed hens. For these labours I received two shillings and sixpence a week and my breakfast each morning, which was the best meal of the day.

I generally worked Saturdays as well to get an extra shilling. I got an egg a day to take home as well. It was hard work but it was worth it all to see the delight on mum's face when I took my little wage home.

It got harder when I was about eleven, as my teacher though I had a chance for Grammar School and so gave me a lot of homework.

I passed the exam but mum and dad said that what was good enough for my brother and sister was good enough for me.

So when I was fourteen, I had to go in the mill. My sister taught me to weave and after a few weeks time, if a weaver was off work I sometimes got to look after two looms. The wage was around about ten to fifteen shillings.

Things had improved in the cotton industry by then as far as hours were concerned.

We worked 7.30 a.m. till 2.00. In those days if you wanted a drink of tea, you had to pay two pence for six hot water tickets. You had to take your own tea and pot and when you went to the boiling water kettle an old man stood there and you had to give him, a ticket or you could not have any hot water. Good old days!

You had to wait for someone to die before you could progress from two to three or four to five and six looms.

Every holiday, I went back to the farm to earn a few extra shillings and when hay time came around I rushed my tea down and ran to the farm to work until dark each night and weekends. I remember in them days, all the people who lived round us were very poor. If anybody decided they had enough each to make broth they might make enough for half a dozen houses. All the neighbours would get a big basinful.

I remember when the general strike was on, I set off one night to an old local tip with a friend or two. We rooted and dug and eventually we filled a sack each with what we thought was coal. I carried and dragged it for about three miles, until I was nearly exhausted when I got home. My dad examined it and immediately dropped it in the ash bin because it was all stone. I really cried.

When I was about twenty-one there were five young chaps of us. All had six looms each in one corner of the shed and we used to get up to all sorts of tricks. One of them would wait till you were bent over a loom and he would fill a paper bag with water and throw it.

He had a good aim. Ninety-nine times out of a hundred, it would burst in your face and when you looked up he was nowhere to be seen.

One day, they had been acting the goat most of the time and when one of them went to one of his looms, there was some orange peel fast in his warp ends which made him a real mess. He was so aggravated that he ran round to Frank's looms and grabbed hold of a bunch of warp ends and tore them out. That was the end of any horseplay for a day or two.

I remember quite vividly in 1925, I was fourteen years old, and one of the sons of the owners of the farm had a Ford car and he took me to Blackpool for the day. If 1 had been taken to America, I could not have been more thrilled.

I never though such places were on this earth.

As days went by, things were slightly improving at home, all except dad's health, and on the day after Boxing day 1927, he passed away.

In the new year, all four of us went to work and we were not long before we had paid all our debts. Before the holidays, one of the young men in the village said he would sell

me his hen coop for 5 pounds which was very cheap.
Through being very careful I had saved 50 shillings, so I put that down and paid the rest off later, with the money I got for haymaking.
I progressed from there and eventually had four acres of ground rented and four hundred hens.
The cotton trade was very up and down in those days. Sometimes we would all be working flat out and then all of a sudden, we were down to two or three looms.
The owners of the mill were very strict and we dare not answer back if we were reprimanded. In those days, every loom that could be packed into a weaving shed was packed in. There were over a thousand looms in those three sheds. It was a very hard life but we also had a lot of fun.
There was no traffic to speak of then. On warm summer nights one or two would play melodeons on the grass verge opposite the houses and scores of young people would dance in the road. And in winter, anyone who owned a piano would invite a few people in for a sing song.
At the mill, it was always "waste time" at about 8.15. This consisted of two of the bosses standing at a table and each weaver had to take their cotton waste and tip it on the table to be examined. If you had a cop bottom, which they considered was slightly bigger that it should be, they made you take it back and weave it up.
There were a few teenagers worked together and one morning unknown to one of them, they put a library book in his waste tin.
You can imagine the bosses' faces when this landed on the table and what they said to him. We had some happy times.
I was Church of England but my friends were Baptists.
I had a decent baritone voice and my friends asked me to join their church choir which was known very well as being very good.
 After a bit of trouble at home, I did this, mum saying she

would rather I went to the Baptists than nowhere at all.

I had my voice trained a little and sang at various chapels and churches in the district. About 1929, the choir decided to give comic opera and each ensuing year until the war we would give one of Gilbert and Sullivan's operas at our own place for a week and then all over Rossendale after.

At the mill, we formed a football team.

As I got a little older about two dozen lads and lasses from the mill went dancing at the Mechanics Institute at Burnley every Saturday night and we had some wonderful times together.

Mum stopped working and got ten shillings pension each week and although money was still not plentiful, it was a hundred times better than before. I was twenty- one when I had my first week's holiday at Morecambe Bay holiday camp and I had a wonderful time.

The only time any relaxation was allowed in the mill was at Christmas when we paid six pence each to have a party.

In 1933, the great crunch came. Mr. Spencer of Burnley decided that everyone he kept in employment should run six looms and make the rest redundant.

There was a strike and a lot of trouble, but eventually, he got his way and where previously you could earn about 3 pounds a week from six looms it dropped to about 2 pounds.

My sister, brother and I were what you might call fortunate, because we were given eighteen looms together. My brother got married, then my sister and eventually in 1937 so did I, to a young lady from Colne. She also was a weaver.

I rented a back-to-back house from the firm - one up and one down. They found her work at the mill. At that time, if a house came empty, if you didn't step in smartly, it might be twelve months before you got chance of another one.

We had a quiet do at th' wife's place. They were chapelers. And then we went to her house for tea and a do at

Beatle or something and that was it. I sold my poultry to buy the furniture.

I was made a loom over looker at the mill. Soon after my wage was increased tremendously.

I remember my first over looker's wage was 4 pounds. Since the day I was married, I've always given her my wage - every penny. We've been short but if we've had a penny, we've joined at it.

I've seen sometimes I've worked overtime and I've thought I won't give her this. She doesn't know but then I come home I've stuck to it and then I've thought she knows. She knows and I've given it to her eventually. We've never kept anything from each other.

We had to carry all our own warps on our shoulders to every loom and on the next set of looms to mine was an old man who was very short of breath. I think he must have had a bad heart, so I used to carry warps for him too – a total of 264 looms. He used to give me an odd packet of cigarettes. He'd been a fairly decent bloke with me. He'd taught me a few things but if he tried to carry a warp he'd been gasping for breath. I was fairly reasonably fit and I carried every warp for him.

There was plenty of work from about 1946 to 1949 and then things started to revert to their old pattern. I moved about from mill to mill and where to earn a wage.

But things were not as bad as they had been in the twenties and thirties by a long way.

The hours in the mill were shortened and we got longer holidays and one of the best things we got was holiday pay. This was paid on a percentage of earnings.

It felt wonderful having money at the various holiday times.

3. OLD HANDS: THE END OF AN ERA

The old hands were those who started work in the Lancashire loom era, before the age of automation, when in the opinion of some, weaving lost its designation as a skilled craft and became merely machine minding. Many of those experienced workers were part-timers in the seventies.

Quite a few had already passed the retiring age and a substantial proportion were married women. This pattern is reflected in the choice of people included in this chapter. Bertha and Harry, Edna Sidebottom and Fred Miller are all over sixty. Mavis and Rose are married women with older children, who manage to do overtime, as well as a full shift. Nellie and Susan have younger children and are able to work a combination of hours, which changes according to their family circumstances.

Reg is one of the immigrant workers on whom the industry was becoming increasingly dependent for skilled and unskilled labour in the seventies.

Reg and his wife, like a number of other couples, manage to take care of their small child and work by being on opposite shifts. One works mornings and the other afternoons.

Those in their thirties, who began work on Lancashire looms are perhaps, as Susan remarks, the last of an era, when girls followed their mothers into the shed and before automation took over.

Change

A recurrent theme of their accounts is change; change in the job - from Lancashire looms to automatic looms; change in the organization of work; change in the relationship between workers and management and workers and workmates; changes in attitudes to work and living standards and the passing away of familiar customs.

Fred Miller notes that the second world war marked a period of drastic change at work, when the intricate balance of rights and duties and responsibilities between apprentices, over-lookers, weavers and management all seemed to disintegrate.

After the war he says things became slipshod; labour had changed and it wasn't the same fun working in the mill as in prewar days. Now he says there is not the same interest in superior quality work that there used to be.

Now the slackness of the forties and fifties has given way to the automation of the sixties and seventies.

The rhythm has changed again. Gone are the days of sunbathing in the summer in slack periods, of taking knitting and library books to work

The weaver is no longer standing in the middle of her four looms but is literally turning round a set of thirty or more automatic looms, to keep them all going at full speed.

Nellie describes it as all speed, concentration and organization, - with no time even to stand and drink a pot of tea in the ten minutes break.

Smokers have to dash to the toilets and back - no time for a chat and a giggle. There used to be more opportunities for enjoyment at work twenty years back, parties, concerts and fun. Now it is just a matter of work.

Observant workers, such as Mavis, see changes taking place now from week to week and wonder what is going to happen quite soon, when those workers over sixty are compelled to retire and cannot be replaced by people with similar skills and motivation.

Not only are the types of recruits into the mill quite different from the prewar days, but also there was formerly a regulated progress for promotion which had to be striven for. I n t h e
s e v e n t i e s newcomers only wait a week or two on battery filling perhaps and then they are on looms. Often as Bertha says it is merely a case of having someone on the set of looms, whether they can weave or not. Whereas formerly would-be apprentice over lookers had to wait for their turn, now there are factories with no apprentices at all. Mavis sums up the position gloomily by saying, *'It's a dying trade. It'll be a slow death but I think it's dying. Other countries have caught up and passed us. They make it (cloth) cheaper."*

An important fact stressed by the women workers is the cooperation between kin - sisters, mothers, daughters, - which makes it possible for mothers of small children to work either full or part-time.

Some work opposite shifts or half a shift with a relative.

So close may be the cooperation between sisters that like Rose they feel as if their sister's children are their own.

They share them in a very real sense, especially if for some time they have lived in the same house or have always lived near to each other. The women compare themselves and their situations with those of their relatives and friends. They have a much easier more affluent time than their mothers had, but they do not intend to let their children follow them into the mill. Even daughters who show a genuine interest in weaving are sidetracked into hairdressing or something else. Many of their sisters and friends are in part-time jobs which are lighter and in more pleasant surroundings, but they are less well paid. Some have chosen to stay at home on a lower standard of living. But those who are working in the mill feel they could not stay at home, though from time to time they may apply for or try other jobs which are less tiring. It is not unusual for women weavers to be earning as much or more than their husbands, and standards of living for weavers in two or three wage families are high, with freshly decorated houses, coloured television sets, processed foods, cars, trips, dinners out and annual family holidays in England or abroad with the children; - a far cry from their childhood when many did not have the coppers to go on the school picnics or outings.

Many of the women too spend considerable amounts of their weekly earnings on hairdressers, clothes and shoes. The ones committed to the job, such as Mavis and Nellie, talk about the atmosphere of mill work which they enjoy and would miss elsewhere. They enjoy the contact with the other workers and above all the gossip about people they know.

It has a hold on you somehow says Nellie. It has a smashing atmosphere she says. It's like a little world of its own.

In Susan's words, *"There's plenty going on and you find out about other people. You don't just go and not speak. You find out about people, about their lives and they find out about yours."*

One significant point which emerges in the accounts of the old hands, as well as the retired, is the attitude to work.

The old and middle-aged were brought up to do it and so accustomed were they to work that some found it difficult to keep away from the shed when retirement came. Work was their life. Nellie perceptively describes this internalization of the need to work. They were brought up to it by their mother.

They were pressed to go to work, even when they were sick and still do. She is sufficiently self-aware now to realize that she pushes herself too hard at work and to partly understand why, but she still feels physically incapable of relaxing and doing less than she knows is possible.

Among the retired it was only Ethel who shared this same kind of objective self awareness and felt ambivalent about her situation. At the same time there is an increasing rejection of the out- dated image of the inarticulate, female weaver in clogs and shawl. Susan complains that when she goes dancing, they say what's your job and they look 'daft' when she says she is a weaver. They don't think there are any left and that if you are a weaver, you must be 'thick'. An important innovation of the era, touched upon here, is the employment of Asian workers.

Remarks are passed that they don't mix very well with the local workers.

One problem is communication.

Each has a problem understanding the other's way of speaking English and in the shed the newcomers are not used to the sign language and lip reading, with which the old hands overcome their deafness caused by the roar of the machines.

Again change is remarked in the new customs supplanting the old - Friday off instead of dressing up in the alley. But perhaps the main impression of change, perhaps revolution, is the fundamental fact that cotton is no longer a way of life, but just one among many possible ways of making a living and that life outside the mill is more vital and important.

There are homes to decorate, holidays to plan, outings to enjoy and above all children to educate, so that they will have an easier, better life than their parents are having now.

For the experienced men and women, who choose to stay on in the mill, it is the work they know best.

The pay compares favorably with other jobs and after all they like the gossip with their workmates, the atmosphere which we shall try to recapture in the last two chapters.

Fred Miller

Fred is a fresh faced, well-built man, who looks younger
and healthier than work-mates of his age, perhaps a result of
his years abroad in a sunnier climate and his present
coastal living.

For he now lives in a sea-side bungalow and commutes to
work three days a week. He says it keeps him trim.

I started work in 1921 at the age of fourteen I've always
worked in cotton. I started full-time straight away- 7.45am
to 6pm. I started as a boy in the weft room issuing weft at
13/- a week. When we got a new manager, I asked if I could
learn to be an over looker. Really it was the over lookers who
used to choose the next apprentices. If they didn't like you,
they wouldn't nominate you.
Anyway the over lookers said I could train to be an over
looker. Bradley's was in advance of other mills at that time. It
had its own day-release schools for you to attend. The
weavers, the designer, the over looker would all teach you.
You got an exceptionally good grounding, but no certificates.
It was such a progressive firm that your experience there stood
you in good stead later on.
I began as an apprentice over looker at seventeen and a half.
You had to pay 2/6 a week for two years for being taught. It
was paid to the over looker. When you got a section of
looms of your own, you paid a tackler at either side of you £1
a week out of your wage, so that you could call them to help
you at any time.
In three months the position was revised.
Then you might keep one tackler helping you for another
three months, and then you might have both knocked off,

if you were managing alright.

It's a trait of over-lookers, that they can be mucked up and have plenty of work on and nobody gives you a lift. Paying over -lookers for their help stopped in the mill in 1939. I don't know if it obtained elsewhere. It might not be everywhere.

In those days, everybody was on piece of work. It was an admirable system. The over-looker had a small standing wage, and it was made up by so much for every pound the weaver on his set earned. At that time, I had forty looms in a set - no plain ones, all jacquards and dobbies. It was an ideal way all round, because the weavers kept you working and you kept them working. It all seemed to go by the board when the war started (1939).

I suffered a bit at first when I was learning to tackle. The new manager, who said I could do it, was a relative of the mill owner, a college boy. He didn't know that it was the tacklers themselves who picked new apprentices. The over lookers were determined he shouldn't tell them who should be an apprentice, so a war started between the tacklers and the management. They wouldn't lend me any tools for a start, so my mother had to buy me a full set. They made me suffer. They used to throw my tools under the looms and all sorts. Anyway, the management was adamant that I should be an over looker but I suffered for the first few months.

I was an over looker for twenty-five years. I was there till I went abroad

After the war in 1946, I saw that labour had changed. It wasn't the same fun working in the mill. They had got all these innovations.

There was a standing wage. Things had gone sloppy.

Then I saw an advert for a weaving manager in South Africa, I got it and went. I was out there for twenty years. Now I'm sixty- five and doing a bit of part-time. There've certainly been a lot of changes. I was brought up in a place where every piece of cloth was guaranteed.

Now nobody is on their toes. Anything goes. Now nobody is interested either in maximum efficiency or quality work. I think it's a common malaise of this country. It is not peculiar to textiles. But you know every generation thinks theirs is the best.

But from what I've seen of the youngsters coming into the mill, I don't think much of them....

Bertha and Harry

Bertha and Harry live in a modernized terrace house with fitted carpets, central heating and other modern conveniences.

Harry's hobbies are Do It Yourself improvements to his house and car. Bertha retired at sixty, but still goes to the mill down the road on three afternoons a week.

She said she enjoys the company now her two children have left home.

Bertha. We went to the mill from fourteen
Harry. Well I started at the Charity Mill, Gingham. I went in learning to weave, though I didn't want to go in weaving. I wanted to be a joiner.
Bertha. But in those days you had no choice.
Harry . Through circumstances like..
Bertha. Like our parents - they wanted the money. You know what I mean.
Harry. It weren't a case of serving your time till you were twenty one..
Bertha. And we had to go in the mill both of us.
Harry. I didn't care for it at all - weaving, but I got used to it. I found out after a bit - I'm going to be stuck with this job, so I might as well try and get to know all I can. So naturally I carried on.
Bertha. You learnt weaving from your dad didn't you. Then when you'd learnt you got on looms.
Harry. That was after three months time. Then in six months I had three. Then at the end of twelve months I'd four. That was quick going for those days. It was a matter of circumstances really. We'd people giving over and they moved the three loom weavers onto the four.

Bertha. They moved them up. And then in them days they were not like they are now. They only wait about a fortnight and they're on looms.

Harry. Oh aye, they are now.

Bertha. I were sixteen when I got two loom.

We'd so long to wait you know and then you decided you'd like to learn tackling didn't you.

Harry. The Charity Mill closed down in 1932.

I were twenty one then. I came standing for work at Thorpe's where Bertha worked.

Bertha. You never got on did you.

Harry. I stood there practically every morning for six months, and I never got one day's work

Bertha. There were nothing at all then were there. The cotton were just hay wire. And then it picked up again didn't it?

Harry. After six months playing, I went to Stone Mill they opened it down at the bottom of Cow Lane. I went on four loom' there.

Bertha. You got on there didn't you? They were just starting up.

Harry. And after two years they decided to go the *more looms* system, which was six looms instead of four, at a reduced rate: more load for the worker for practically the same money. Everything went straightforward. They got rid of some weavers. They just sorted the wheat from the chaff. They didn't bother about sacking then.

Bertha. They picked the best you see. But it's not the same now.

They'll have anybody now - just having somebody on the looms whether they can weave or whether they can't.

Harry. I worked for about two years and then they made me an apprentice tackler.

I was always playing about with my looms.

I used to stop in after the mill had stopped and started playing about with them, and the manager caught me doing it. So he asked me, he said, "*would you like to learn tackling*".

"Certainly I would," I said. l were playing about with them trying to find out what I could.

Bertha. One way to get his work out wasn't it.

Harry. Oh I enjoyed it.

Bertha. Anyway, he made a good one he did. Let's put it that way didn't you?

Harry. Better ask my weavers. They always say that self praise is no recommendation.

Bertha. But folk talk don't they, that we know.

Harry. It took me eighteen months as an apprentice. I just did eighteen months.

 Bertha. That was good wasn't it? That was good for them days.

Harry. There'd be a tackler off, and I'd go on his looms.

 I had   to go with another over looker and he taught me all he knew. He was a good tackler, I worked part time with h i m on his set.

Bertha. And then he had a set of 'is own, a little set.

Harry. The apprentice set - sixty looms. Now they only have eighty.   Then they had ninety-two.

I had sixty against ninety-two in those days. The wage on a full section was about £3/15.

 Bertha. Yes its true is that.

Harry. And then we had to make us own wages. We got paid off what the weavers earned.

Bertha. It weren't like a standing wage like they get now at our place. You got like ... If the weaver did well, then the tackler did well you see. But if you had a rotten lot of weavers then the wage were down.

Harry. Well, I averaged about £3/15. Sometimes it was as much as £3/18. Three shillings were a lot more. It were a lot then.

Bertha. Weren't it? Eh, we thought it were a decent wage then:

Harry. An over looker earned twice as much as a weaver.

Bertha. Because when you think when I were learning, I were sixteen when I got two loom', and it took me all my time to get a pound.

Harry . Yes we'd two looms

Bertha. Yes, it did! And I only had three loom' up to getting married hadn't I? I weren't married till I were twenty-five. Then we went on six looms, I were only working twelve months and then I'd to give over because I were having our Frank. I stopped work ten weeks before he were born. Our Mary were only five weeks old when I went back to work. Because you couldn't afford to stop at home in them days. I had six looms then didn't I? and you were on a full set.

Harry. We were renting a house down Albert Road then.

Bertha. We came in this house in 1942. We've been in this house ever since. I've kept wanting Harry to move but he's stabilized. He won't move.

Harry. There's no point in moving if you're happy. After eighteen months I went on a full set. It was the management that got me tackling. There was another young fellow learning to tackle at the same time as I was and his father was a tackler. And in those days if your father was an over looker, the son got in .

Bertha. Yes it was a case of if the father were an over looker the son had to be, you see. More or less, they got the first chance didn't they?

Harry. Anyway, the manager must have thought I was more capable of running them then the other lad, because he put me on first, and the union objected.

But the manager stuck out for me and the union gave in at the finish, and I got in at twenty-six.

Bertha. You were the best of the two.

Harry. The other lad had to wait till I'd finished. They'd only learn one at once. They only had one apprentice set, and then he followed on the apprentice set behind me.

Bertha. Oh, then most tacklers were the sons of tacklers. If your father were in they thought, oh that's his son, we'll have to get him in as well- even if he weren't really ready, which I didn't call fair. It was just a case .....

Harry . Oh, it was a closed shop was the over lookers' union.

There's no apprentices at the factory now, not one.

In 1939, when the war broke out we were a reserved occupation and I carried on overlooking till 1941.

Bertha. And then you had to go for a medical.

Harry. I'd to register, and I registered for the navy. Anyway, I got sent to an aircraft factory an aero engines I had to fix - the same job. It was right monotonous. Eventually, I came back home and started work at Black Lane. I got a fortnight's work at Black Lane.

Bertha. It was like bad to get after the war till things got sorted out.

Harry. I did two weeks there. The union sent me there, and then a man came back and I was out of work again so I went back to the union and they sent me to Prosperity Hill.

There they had various sorts of looms, circulars, plains, dobbies. I was just there three weeks.

Bertha. And then you were sent to Thistlewaites'.

Harry. Then the over-looker came back there, and I was out of work again.

Bertha. This old mill was still closed down then. It hadn't started up.

Harry. And then I went to another place, I'd just worked there nine weeks and then they opened my old mill, and they put Jaquard looms in. They weave like tapestry and brocade. They asked me to go and start them up. I was tackling those looms for four years. In fact, I got them all going.

There were only six set up when I got there. I found that work right interesting – always something different.

I'd been there about four years when the manager left to start a mill of his own, so I got his job as manager.

After another year, I decided to branch out on us own.

Bertha. You started up on your own didn't you?

Harry. Soon after I'd left that mill closed down I didn't like being on one job all the time.

I thought I'd improve myself, which I was doing, - to get a step ahead.

I got forty looms. I went in partnership with Bob Illingworth. He was a friend.

Bertha. You knew him well.

Harry. We bought the looms from Baileys' at Colne. They were selling them off cheap. There were old looms but they were good ones.

I got forty for £230. You can't get looms for that price today.

Bertha. Now those new automatics in the shed cost several thousand pounds.

Harry. We started to weave air cloth interlinings.

Bertha. And then there was a slump, wasn't there?

Harry. We had to sell all the lot of them on block. We did very well while it was running. We were making more money than tackling. We managed to keep going from 1951 to 1955. By the third year we were doing very well. Then the slump came and we had to give up. We couldn't get orders.

Bertha. You learnt a lot of weavers didn't you.

Harry. We had seven weavers. They were getting eleven percent above what they would have got elsewhere. They were earning about £7/12 shillings which was a good wage then.

Bertha. I just helped at first and then I started weaving and got my own wage.

Harry. We got an old shed in Trumpton. It was an old mill that had been emptied before the war. A chap had battery hens in it at first and then he used it as a small engineering shop an' then we took the bottom half. The rent was only £3 a week.

Bertha. It was a cold place.

Harry. It was, till we put heating in.

We did all the work ourselves - cut-looking, book-keeping, tax returns, typing and overlooking.

Bertha. If they had have let us run another two or three years we'd have been well away.

Harry. We couldn't get orders. We had to stop the looms.

Bertha. The weavers were right upset when it stopped.

Harry. Two of them cried.

Bertha. They really liked it, and they were near home. They seem to come and leave these days.

Harry. A lot of over-lookers have gone into different jobs.

Bertha. You wonder why the young ones won't come in.

Harry. It's been slave trade ever since it started, and the poor weaver's been the bottom dog.

Bertha. Weavers never strike.

This afternoon … them eight - six were all stuck together. They wouldn't go at all.

Harry. It's because there was no humidity. Those beams are sized for a certain temperature and humidity.

Bertha. It got four o'clock and all the looms round me were stopped. It were shocking, everybody were complaining. The manager's a poor do.

Harry. I've forgot more than he knows. How many jobs has he made right since he went. I've been there seven years, and I've never seen him make a job right yet.

Bertha. He's having a week off next week. That place 'll run smooth then. He has a watch and he's looking at it all the time, pointing at it if you are half a minute late.

Harry. He picks on the wrong people - people who never do anything, he picks on them.

Bertha. One young woman asked for half a day off, he said "You've just had a week off'.

And she had it for her husband's funeral! He is silly. He's comical.

Harry. As soon as he opens his mouth, he gets under your skin. He's no tact at all.

Bertha. He'll happen be better when he's had a week off.

Harry. He'll be no better.

Mavis Atkinson

Mavis Atkinson lives in a freshly decorated, semi-detached house, - pebble-dashed, with a small garden. Her L shaped lounge-dining room is strewn with nylon rugs.

Her display cabinet is overflowing with china and pottery animals.

On the walls, prepared in several shades of green, hang painted plaques and seaside scenes. A large, coloured television stands in one corner of the room.

As she talks, her husband quietly sucks his pipe and watches a football match.

My mother was like I am today. She tried to keep me out of the mill, like I tried to keep my children out. I never found out why she did that. She got me various jobs you know. One was at the toffee works. I came up with all sorts of excuses and she kept sending me back and I kept coming home and then I went back again, and then she found work in service and I didn't like that.

That was less pay than in the mill in those days, nearly forty years ago. She finally let me go in and I was as happy as anything then.

It were what I really wanted to do. I had an older sister that worked in the mill. Why my mother didn't want me to go in, I don't know.

Funnily enough I never asked her either.

I know she tried to keep me out and I wanted to go in. Mi mother and father were both weavers and I learnt to weave wi' mi mother.

I learnt to weave for about four months and then we

were put tenting. I remember I got 6 shillings and 9 pence a week wage and I got the 9d. We didn't get paid to learn in those days, and mi mother used to let me go home for mi breakfast.

She didn't go home, she had half an hour and I had half an hour. And then I'd go home about eleven thirty and make the dinner, and then I'd go about four o'clock and make the tea. I used to hate having to go home. I kept weaving till I got married and then I left and went working at Merryweather's. I didn't stay there very long, because I went working in a shop during the war. It was less pay, but the hours were more convenient, as I had my little boy then and he hadn't started school.

Since the war, I've been back weaving and left several times. At one time, we went to live in Birmingham for two years, and then we came back and I went back to weaving to Snodgrass and Marks's.

I was there about twelve months and then it closed, and I went to Winkles'. They closed that and I went to Summer Field to Beavers' and they closed that one, and after I left Beavers'. I went to William Ridges' and they closed that one, and I went from that to another of their mills at Colne and that closed, and I left and went to Millsons' at Nelson and I'd just been there three months when they went on short time, so I ended up back at Millers'.

I were so sick of the redundancies and short time working. Then I got involved in union activities. I got involved because nobody else would do it.

They balloted round th' mill for a representative and they put my name forward and I was elected. As well as being mill representative, I collect the subs and from there I went on to th' committee and I stayed on th' committee for about two years and when I went to Millers, I started working Saturday morning and they sacked some off the committee. They said I was going against all the principles of the union.

That's what they told me for working on Saturday mornings. So they asked me to resign and I refused, so they sacked me.

I liked union work and all, I enjoyed it. I enjoyed getting involved with the demonstrations. I liked going to London. We went to the Houses of Parliament and we were meeting the MPs and arguing with them.

That was only about three or four years ago.

We went on one demonstration about the Viet Nam war. I went a few times. One was about the Industrial Relations Act. That was the main one, and there were the normal conferences. They came up with all sorts of issues that never got any further than the conference table. All the time that I was there they were talking about trying to get weavers superannuated but that was as far as its ever got, - round the table, and a shorter working week. You know, more time off, more pay, and less hours.

They're getting somewhere with the extra holidays. We've got three extra days this year, but I can't see them getting a shorter working week.

They might get down to thirty-five hours, but I can't see them getting any shorter than that.

The wages in cotton are low in comparison with silk.

I came out of silk into cotton and then three years ago, the wages in silk were five to six pound a week more. Nearly all the silk mills round here have closed down, so we were tied to go back in cotton or travel.

A few are managing to get £40 and more these days but it's unusual for a weaver to earn so much. The rest of them are nowhere near that.

They can get it with overtime though.

Nearly anyone can get that with overtime, but for a week's wage, I think it's exceptional.

I think it's marvelous.

I don't know why weavers fell behind in wages because

even before the war their wages compared to any man's
wage, because wages were very small then.

When I had six loom' when I got married, I'd three pound
something. I know if I had over three pound, I were laughin'.
I never let it get under that £3.

It always had to be over that £3.

But even then I'd more wage than my husband, so they must
have compared to men's wages even in them days.

I think they all get more than their husbands now, unless
they're shift workers or pit workers. I don't think they do
(get more) than night workers. But an ordinary man's day
work, - I don't think they're earning any more than a weaver.
All my life as long as I can remember I've earned more than
my husband, up to recently anyway.

It's not just one person I think everybody used to. At that
time there used to be quite a few men in weaving but there isn't
now.

There's never been one rate for men and another for women.
The weavers have never been on strike really.

That six loom strike fizzled out quick. I remember my
mother said she went in.

They were all standing outside the factory and she went in 'cos
she had to do. My mother couldn't afford to strike.

They all went on six looms at the finish.

There were tenters, two, four, and six loom, weavers and you
gradually went up the scale. I had six loom' by the time I were
twenty.

You'd to watch it then, there were loads of young ones there
then, and you'd to watch it, that one didn't go before you. I
were always in that warehouse.

You used to have to go to the book-keeper then and I'd say,
*"You've put so and so on two loom' and I came before her I came in
September and she didn't come while Christmas." "Oh",* he'd say,
"You'll be the next on."

He must have missed or they tried to put you near your

family, near your mother or sister - somebody to keep their eye on you.

It was the book-keeper who moved you then. The shed manager we had, like I said, you couldn't talk to him He just walked round twice a week and he didn't know the first thing about weaving.

But we hadn't foremen. It were the boss tackler what we had then... he got managing later. He were the foreman, except that he had a set of looms to run as well, but there were nobody only this book-keeper as we'd to go to.

He used to have the wages to reckon up as well. He used to reckon 'em up and them in the office 'd put 'em out. If it were a copper or two wrong you'd go back to him And it were him that allocated the looms as well.

We used to have a system. If you had a float or a smash you'd to have a ticket on it with somebody's initials on it, and then you could let it go and you used to have to go to the foreman or the book-keeper and if one wouldn't give it, you'd go to t' other one.

It had to be somebody official, so long as it had their name on you could let it go. We used to play 'em off against one another.

They'd fine you six pence for a float, if you hadn't a ticket on it, and you'd just let it go, you'd said, "*Oh blow, let it go*".

They stopped it out of your wage for probably this yard of cloth that you'd spoiled and then at the Monday morning we'd go in the warehouse to buy a fent.

We used to wear 'em tied round and they'd sell you a piece back. Then you had to buy 'em back. They'd fine you for it, and then they'd sell it to you.

They'd charge you 3d then for a fent. Comical isn't it! You used to be able to buy cloth by the pound then, and pillow cases and sheets cheap.

I don't know why they don't do it now because they must get it returned. Some factories do still have shops. It's the system they introduced which has put the weavers' wages up now.

Before they were getting £30, £32, £34 and then they went up with a bang.

Oh they were delighted. I can't see it going back, because it's working very well and I don't think anybody complains now about these wages.

There are a few grumblers now and then but I don't think they're really serious about it. It's just something to natter about.

But I think like they say, it's a dying trade. It'll be a slow death, but I think its dying. They'll always need cloth, but other countries can make it cheaper, can't they?

Other countries have caught up and passed us. They make it cheaper.

There's very few weaving mills left now.

They've closed so many.

All my family were weavers, my sisters, my mother, and my father and everyone of 'em got out of it except me. And they think I'm crackers for staying there.

One sister works in the office at an engineering place, as a clerk and my other sister is a barmaid.

They don't get as much money at all but they prefer it, as it's an easier living.

My older sister said you get less money but you adjust your standard of living. You finally get used to living with less money.

My husband calls me greedy.

I've never wanted to do anything else. The times I've gone out of it... there's something about mill life I don't know what... It's working with so many people probably. It's that I like, - the contact with them.

The few times I've gone out of it, I've missed it. I've always gone back.

I think about my sister being nice and clean and working

in a bar, you know, I mean she's plenty of people to talk to hasn't she?

I like mill life I always have. I'll probably stay there till I retire. I never meet my work mates outside. I might see them when I'm out shopping, but there's very little social activities go on at the mill.

I don't know why there's such a shortage of labour. Where are all these weavers from all these mills that's closed down? They had hundreds and hundreds of weavers.

They haven't all settled down in fresh jobs.

There must be hundreds and hundreds of weavers and all the mills they closed down in Lancashire. There must be. They haven't all died off.

Simon Plunge closed down three mills in four years. All of them haven't died off. That was about four years ago.

There must have been three or four hundred weavers.

There were nearly a hundred at one I went to that closed.

All the departments are short these days...preparation, department, the warehouse, the weaving shed, the winding. There's not a lot you can do to improve the conditions in a weaving mill to attract more people. The only thing you can do in a weaving mill is to keep it cleaner and they are being kept cleaner.

But you see the wages in other industries now are getting close to textile wages. I think that's one of the main reasons. Women can earn about £34 a week in engineering, so they're not going to come back into textiles.

I don't know whether it's harder work or not. I've never been in engineering.

The majority of weavers prefer shift-work, and in shift-work, you can only do six to two, and two to ten.

All they can do is shorten the hours. The night shift six to ten p.m is mainly for young mothers with little children, but I wouldn't like to work days.

One thing I like about shift work is that you have part of

every day at home. I wouldn't like to go back on to day work. By the time you get home, you do your housework at night, you do your washing at night. I think that's more bed and work than shift work.

My husband helps me but there's a lot don't.

I think that's one of the reasons why one of my sisters left, because she didn't get any help at home, and then it is hard work if you don't get help in the house.

I think once they get adjusted to shift work, the majority of them like it.

I know we soon kick against days if they mention it and then its shorter hours on shifts 7.30 to 4.30. You've to get up at 6.30 just the same and you're only home at 5 o'clock and then you've a meal to make.

It depends who's home first in our house...who makes the tea.

And I think that's the way it should be. If Fred's first he makes it. If I'm first I do it. The one that's in does the work.

Newcomers? The type of girls coming in now don't stay. They're nothing like the old weavers and I think we're responsible to a great extent...the parents are responsible. Because we all have that feeling that we don't want our children to go in the mill. I always say I worked there to keep them out and I think it's a mistake.

I have one daughter and she'd give her eye teeth to go in just like I were when I were young. And it's only recently that I've seen it.

Both my daughters are hairdressers. My oldest daughter asked me several times to learn her to weave but I wouldn't listen to it. They both went to the High School. I did just like my mother did. And yet it didn't dawn on me. And I've thought since, why should I have kept her out when she wanted to weave. One reason was probably that trade was bad. But I think we have done a lot of it, because you can ask any weaver if she'd put her children in and she'll say, "No"

Nine out of ten won't put their children in. Because you want your children to have an easier life than you have had yourself.

But of course today they're not weavers are they? They're loom minders.

I don't think there are any real weavers, unless they've been Lancashire loom weavers, - a Lancashire loom weaver who can do a good basic grounding. It's only Lancashire loom weavers who can do anything.

For instance, as far as weaver helping is concerned, I don't think they make a good weavers help, unless they've been a Lancashire loom weaver.

There's so few young ones coming in today. You can't compare them with the old ones. The young ones that come in, they go battery filling and there's very few of them are interested in learning to weave.

We've had one or two and they've learnt for a week or two, and then they've gone back to battery filling.

I tried to talk to one of them. I said once she'd learnt to weave she'd be right. She'd be able to weave anywhere.

Once you've learnt to weave at a place like ours, you can weave anywhere. But no, she wanted to be a battery filler.

I should think that when they go from battery filling to learning to weave they'll go on a lower wage for a time. Whether that has anything to do with it, I don't know.

They come as battery fillers and they see the amount of work a weaver has to do and they think that, "*Oh that's not for me. I'll just sail round filling in my batteries*".

Because there's Mrs. Ship she's a battery filler. She went in for weaving, but they nearly all do a week or two battery filling first, and she were quite happy to continue doing that when she saw weaving was hard work.

I know one thing, if there were any weaving down Cornwall I'd go. I'd like to retire to Cornwall. I'd retire a lot sooner if I could find a mill down there.

I think Lancashire's depressing,- the weather I mean. I don't think you ever get used to it. It's shocking. I was Lancashire born and bred, but I can't get used to it.

It's surprising what a lift a nice day gives you. We've always gone every year for our holidays. We went for ten years (to Cornwall) and then we went to Spain. That was lovely weather there.

I did apply to go working abroad. I sent for all the paraphernalia. I saw an advert' for a mill in Australia. But it came to nothing. I liked the sun. They did send us a brochure. It was a beautiful place. They wanted whole families of weavers. They said we'd hear in a few days. That was three years since. But you know you're not forced to settle if you go abroad.

So many come back. I don't think I'd have settled really. I couldn't even settle in Birmingham and we were in a right nice part. I wanted to come back. Fancy wanting to come back to this place!

Of course, it was my family I wanted to come back to. We have a family that won't go anywhere without the rest of them.

There always used to be sixteen of us on holiday together. Where one goes we all go. We all go everywhere together. I know when we two were in Birmingham, my mother said it was the most miserable two years.

It used to be smashing at work for parties and that, but not now. When a lot of these older end have gone, I don't think there'll be a lot of atmosphere about it, because all these younger ones that are coming in...they haven't got the same ...they don't mix like the older ones do. They keep themselves in groups. It used to be smashing for parties in th' shed. We used to dress up and that.

At Christmas time they used to go round the' shed on a truck in fancy dress. They had their faces blacked and striped pyjamas on -like golliwogs- all dressed up. One came with

rags in her hair. They were bits of fent she'd tied them all up.
We used to have a marvelous time at Christmas.
We used to have a factory wedding. It was an annual affair.
One would dress up as a bride and another as a groom. Then
there'd be a minister and every year they'd have this wedding
and they used to tie her to a pole in the shed. We used to trim
th' alley wi' streamers.
We went to Blackpool at th' fifty year anniversary o' th'
mill an' all. At Christmas we always used to have mistletoe
and be chasing each other and have pies and cakes. We
bought our own like, but even that's fading out. Since the
automatics took over you haven't time. You'd have to stop
your looms to do it. We did stop but we'd only four. We
used to have parties round the looms and the managing
director would come round and have a drink.
We had the looms stopped and we were in groups, and
they had drinks and he'd come round and have a drink.
But then when automatics took over that were all stopped.
We tried to have a party this last Christmas but you don't
enjoy it when you have to keep working when your
looms are on. All that's had to go.
We asked to stop for half an hour for a party but they
said we'd have to make it up and then everybody were
mad 'cos the bosses and foremen had a party in the office.
It were maddening that were.
Last year they tied a girl up at work, - Rita, to a pole
outside the shed when she was going to get married. They
did it about a week before the wedding, just outside the
shed door. And the bridegroom weren't there and all.
He were hiding. They didn't catch him. He ran right up
the hill and he were watching and when they all went in
he came and untied here.
He stopped there till they all went in about half past six
after break and came and untied her. They intended
leaving her there and not letting her go back to work.

Rose Swindles

Rose is in her early forties and lives in a six-roomed council house at the top of a hill overlooking a valley of mill chimneys. Her garden is neat but flowerless. Her glass porch is decorated with a cage of stuffed birds.

All available flat surfaces in her front room are filled with models of vintage cars and armored vehicles.

Three trophies for ten pin bowling stand on the mantelpiece. Her husband's leather swivel chair faces the coloured television set.

I left school at fourteen. I had a fractured skull in the holidays and couldn't go to work for six months. I started at a mill at Fence. Mi mother just said, "*You have to leave and go i' th' mill an' then you'll have a trade.*" Mi mother's friend said she'd take me and learn me, but I didn't really like her, and I left after two weeks.
It was war time then and I had to go before a board. They let me leave to go and work nearer home, as I said mi mother were sick.
I stopped at home for a few weeks and then went to a mill at Saleham. I learnt to weave there with a man. He was a bit strict but he was better than the other woman.
She were right moody, you never knew whether she were going to speak or not.
I got put on two looms at first, and then three. I think nearly all mi school mates went in th' mill.
We lived at Saleham then and there were nothing else but weaving.

At th' mill I went to, they used to have three sheds and three doors to go in and when it got to 7. a.m, they closed two and the manager stopped at the third and kept pointing at his watch if you were late.

The over lookers were pretty awkward as well. I got on the wrong side of two. There was old Tom and young Tom, his son, an' one day, I went in th' warehouse and called for 'Old Tom and he said, *"Who told you to call me that?"*

It was a different place from the one I'm working at now. It was all fine wefts. It took about three weeks to weave a piece on them.

You went for your wage in th' office. The money was lying on a board with squares on with the loom numbers written on it.

You were supposed to pick your loose ends off. I was threatened a few times that I'd have to pick 'em off i' th' warehouse.

We used to have to work Saturday morning in them days.
I left there when I was seventeen.

Mi dad had got a job at Apple Tree Mill at Brougham as a drawer-in. He got this job and he were in charge of all th' drawing room.

I were seventeen when I went there. I had six looms. Mi sister were looking' after me mother. Mi brother were at school.

Mi mother died after we were there eight months and then me sister came to learn.

It were that right bad winter (1947) and we used to have to walk to work thro' th' fields. Mi mum were in t' mill before she were ill and then th' one where she were workin' closed down. Then she went cleanin'.

Mi cousins all went weavin' as well, all th' girls.

Mi brother did n't like th' mill. He went in a garage and then he went in th' Navy th' week after he were sixteen. His wife worked i' th' mill till two years ago. They had a daughter.

She didn't go i' th' mill she went in a shop.

I don't know why th' young ones don't want to go in. It used to be grand i' th' country up at Brougham. There were fields across from th' mill and we used to take us dinner - sandwiches - and sit in th' fields.

We had a right good time. And I was in a netball team. We had a league. Every mill at Saleham used to have a netball team then.

Mi cousin went to 'th mill at Saleham. She absolutely hated it. I think she'd rather starve than go back in the mill. She said there's none of mine (children) going in th' mill. She used to dread it every morning. It was her mother who made her go. She stopped as soon as she got married. She said I don't care how much they're earning I'll never go again.

I don't know whether I would have preferred a different job from weaving. I groused every day about something, but I wouldn't complain now. I only once did another job, for three weeks, and I didn't like it. I would be 21 or 22 then.

Just before we left Brougham we used to work three days and play two days. When we left there, me and mi' sister went working at Foxton. I didn't stay long, about three months. Mi sister stayed eight years.

I met Sid in the mill. He was an electrician and came to work i' th' shed fixing some fans for a few months. For the first six months after we got married we used to live with his aunt who had a cafe in Blackburn.

I'd an argument with mi step mother and left home, so we couldn't have stayed there. At his aunty's I used to work behind the counter at night. I could manage that an' I used to quite like it. At first, I used to travel to Solden to work for three days a week but then I got a job at a mill in Blackburn, Turners'.

I had five looms there. It was then we moved to this house. When we'd been married eighteen months, David was born and I gave over at work. I stopped at home with him. But when he was fifteen months old, a lady on the next street said that she would look after him.

On the Monday I went to Turners'.

They said they didn't want anybody. I went to see at Thompson's. You had to go down hill like a tunnel into th' shed. I told my husband I didn't like it.

He said "*Well don't go again then*". So I went to the slipper works for three weeks.

Then the manager of Turners' saw me, and said I could start the next day. He had already come to look for me three times. So I finished at the slipper works on Monday, and started at Turners on the Tuesday. I used to set off at half past six in the morning. I took David from being fifteen months to two years two months to th' woman in th' next street- After that I used to take him to Stonewood nursery. That used to open from half past six in the morning till six o'clock at night. It's just along Fermoor Road.
He used to get his breakfast, dinner and tea there and I paid fifteen shillings a week. There used to be three children at the nursery who went at that time in the morning.
I used to get home with him about six. It used to be rough in winter taking him in the wet and snow but it was alright in summer.
When he went at first they didn't have many things to play with and he used to cry sometimes in th' morning, but later on, they got more toys and he liked it. He went till he was five and then he left and started school. After that my sister came to live with us.

She was expecting and she wasn't married and my stepmother said she should have it adopted, and shouldn't bring it home.

But mi sister wanted to keep it. So I told her to come and live with us. I told her we'd help her. She worked till the December at Turners and then she stopped off and had a baby boy.

For three months after he started school, David used to go to a lady up the street till I came home from work. Then from December my sister looked after him when he came home. I was working on days then.

When mi sister's baby was six months old, I got her a job at Turner's from six to ten at night. So she went. She had to learn how to run automatics. Then I went and asked to learn on automatics.

So then we went on opposite shifts. We did that a while. Her boy Gareth would be about five when she got married and moved into the next street. We still worked on opposite shifts and our David went to her house when I was at work, and their Gareth came to our house when she was at work.

There were only three sets of automatic looms at Turners. There were three weavers, a sweeper, a tackler and a battery filler on each shift.

The shed was upstairs and after the 2 to 10 shift you used to have to go through the shed downstairs. There were only pilot lights on and they said the shed were full of rats and mice. In th' morning th' tackler always used to come late for th' 6 to 2 shift and we'd be stood waiting outside th' door for it to be unlocked. Th' other weavers going past used to ask us what pictures we were queuing up to see.

Th' tacklers used to go in th' rest room downstairs, and we used to have to put us own fires out. We had to pull a switch to stop all the looms. Then the tacklers came up after we'd put it out. You had to do everything yourself; put your own smashes in. (A smash was a large breakage of warp yarn that had to be redrawn, also referred to as a trap and the person putting it in was a trapper).

Two weavers had 21 looms and I had 22. By the time I left there I got a standard wage of £12/5. With bonus it ran to about £13/5.

Before that tackler what were always coming late, we' ad one right bad tempered, what were always shouting and balling at you.

At the place I'm at now, if there's anything wrong with a loom, you have to write it down on a paper. There, you had to go and tell him, an' if he were in a bad mood, he used to just walk away when you were talking to him. Sometimes, he'd even throw a shuttle and shout and swear.

One day all the weavers went to tell the manager.

He lasted for about two weeks after that and then left.

Quite a few weavers used to cry buckets of tears over him. My husband wanted to go down and see him after I came home crying one night. I said they'd laugh at me but he said he didn't care.

One other weaver's husband went down to th' mill to see him and said he would thump him, but they let him see him.

I have met some queer folk in th' mill.

There were one battery filler, a woman at Turners'. Once we'd all gone downstairs to the toilets for a smoke and there were only her and a weaver upstairs and she came down and told th' tacklers she would never stop up there on her own with her again. She said she got hold of her but she never would say what she did at her.

We used to brew us own tea and everyday that same woman used to drink th' top off every bottle of milk.

We never used to see th' manager at that place. He used to walk round th' shed about once a fortnight. Nobody really knew whether you were there or not - only the other weavers. The shed had right big doors and we used to open 'em when there weren't much breeze. If we got straight we used to nip outside.

We had one tackler who used to sit outside a lot, Cyril Woodcock. He used to sit there knitting outside, pullovers, jumpers all sorts. He died about two years ago.

We didn't have a right lot to do then. We used take knitting or read a bit if we were straight. You could let anything go (faults in the cloth). They never bothered.

Leaving there and going to Felton's were a different thing altogether. A lot of people who left there when it closed down in 1965 couldn't stop when they went somewhere else an' had to knuckle to.

I know people who've had loads of jobs since then. Turners' was a grand place to work at, but if anything went wrong, the wrong person would get blamed for it.

I've been working at Felton's for eight years this week. You can't please yourself there, and they have different sorts (of weft) an' that, I mean at Felton's th' tacklers come round every week (looking at the shuttles).

At Turner's they only used to come if you called. At Feltons' you have to write what's wrong down on a paper and we have weavers' helps. There was a little chap as swept the alleys at Turner's. You used to have to put a chalk mark on th' floor when he'd swept one alley. If you didn't put a chalk mark he'd sweep the same alley twice. I remember I once dropped a bit of cotton on the floor after he 'd swept, he was mad at me.

The wages have gone up a lot since I got £13/5. Things have changed a lot at Felton's since I went there in 1965.

When I went at first, the helper weavers used to change round in the middle of the week. Now they stay a full week. Because he worked for a few months in a cotton mill, Sid thinks he knows what it's like. He thinks because the weavers didn't have much to do there that they're all the same. He doesn't realize how hard it is at Felton's.

I have 30 automatic looms. You used to have to sweep your own looms and oil 'em. Th' weavers used to pay th' young ones to sweep for them . I used to do it at Saleham.

You could sweep round the loom when it was going or sweep it at breakfast time. You had to leave go of your brush if it got fast when you were sweeping the wheel. Otherwise, you might break the brush or hurt your hand. You'd to stop the loom to sweep under the boxes.

Anybody expecting (a baby) used to ask somebody else to sweep for her.

When I was about 16 at Saleham, I used to help the girl next to me, I went for her weft, swept her looms and went for the tackler for her. She felt embarrassed going in the little room where the tacklers sat.

You couldn't lift pieces out on th' roller when you're expecting, you have to get somebody to do it for you. People vary in how soon they stop work when they're expecting, I stopped at seven months, but I felt fit. I was bored to tears when I stopped off. Well you can't go for walks on your own and we'd just moved here and I didn't know anybody. You get odd ones working right up to th' time. Those who mar themselves, or aren't so well, might stop early.

In Saleham, I remember we used to have to take our own pieces into th' warehouse. We had to pull them off the roller. At Turners', you lifted th' roller out. But people do help each other.

I remember when I first went on automatics there was one woman, she'd be nearly sixty she only had one lung and th' young woman on th' next set used to lift her rollers out for her on 36 looms. I've never thought of stopping at home and not going out to work.

At first, mi' husband hadn't a right good wage. Now he has a good wage and mi son has a good wage too, but I don't want to stop at home, though I don't really need to go.

I can count the times I've ever stopped off. But I don't just go for company. Anybody as tells you that they're not telling th' truth. They could go somewhere better than th' mill, if all as they wanted were company. It's a bit of money you go for. You can save up and spend a bit more if you go out to work. Once I went to work with a boil. I went to work till breakfast, went to have it lanced, and was back at work by eleven.

If I'm drawing th' National Health, I don't feel like I've a right to it. If you're ill, you can work the sickness off rather than sitting and moping about it.

If I stopped at home for a while I think I'd go mad. I'd talk to the four walls. After David was born, I didn't only go back to work for the money. You get a bit fed up if you can't go out. I wouldn't like to do anything else.

At the moment I'm doing overtime. This week I go 6 a.m to 2 and then I go back at 6.30 till 9.30. Nearly everybody what goes on overtime goes those hours. I can save a bit. I earn about £32 a week normally and up to £38 when I'm on overtime. So we can get extras. We can run a car. I spend a lot more on food. I don't have to scrimp and save. I buy plenty of clothes. Sometimes, I look like a ragbag at work I know, but good clothes get spoilt at work.

David - he's always wanting something, records and things. He works at the same place as his dad. He used to work at Bill Segs shop down Pie street but he used to be riggin' aerials all the time. He tends to have a bad chest so he went working with his dad inside.

We always go on us holidays for two weeks in July- not abroad though. Sid doesn't fancy it. And we go off a lot on Sundays in the car for the day.

Sid used to be chairman of Shortridge Labour club for twelve years. He used to go out six nights a week and he spent a lot of money that way. Then he started only going out one night a week. Now he doesn't bother. When you're working if you see something you like in a shop, you can buy it. We buy a lot of records. David has a cassette recorder and radio, they all work on his stereo. He spent a good bit on that. David used to have his own car. Now he drives ours. He's usually off every night in it. Insurance is dear for under twenty-fives so it puts him off getting another car of his own. They've spent over £200 in the last few months on that car. He had it sprayed and the insurance and new tyres etc. Sid's always grumbling about the cost of the car. He uses it for going to work. Neither of them like walking.

The housework? I neglect it sometimes.

Sid and David don't help unless they're forced.

If I'm ill Sid 'll give a hand.

Last Saturday morning, I worked and he was at home so he got up an' tidied up.

On Monday when I was at work, he washed up after supper. But I have to ask him, he doesn't offer. The windows outside would get dirty for two years and he wouldn't bother.

Those models on the mantelpiece and on top of the cupboard? Sid makes them. He has a right lot of patience with anything like that. Our David hasn't. He takes after me. Hobbies? I like a bit of knitting. An' I go to bingo two nights a week - Saturday and Sunday.

I go with two friends from Saleham. Sunday night costs £1/5. Saturday is dearer. We have six goes on the one armed bandit and then we buy four books (of bingo tickets). I only go twice a week. Some go every day. I don't know how they can afford it. One woman up th' street, she goes every afternoon and evening. She must be spending up to £4 a day. I don't know how they can afford it. I won't be going this Sunday as it's my birthday on Friday and mi' husband 'll be taking me out all day Sunday, and out for a meal somewhere.

David spends a lot of his spare time going on car rallies. He goes nearly every weekend

When we came to live here at first, I had to stand in the warehouse for work twice. That was twenty-one years ago. Once I got work and once I was sent home. Then it was the manager picking who he wanted.

Mi mum said in her day it used to be the tacklers in the warehouse, an' she said it were a question of if your face fit. If you were one who used to soft soap all th' tacklers up you were alright.

She used to have to wait in t' warehouse for work a lot in her day. We used to get paid off pieces then. Sometimes if you had a lot of pieces that week, you used to take 'em off and put 'em up on t' racks and save 'em for t' following week so you wouldn't have a good wage one week and a poor wage th' next week. Pieces for one week had to be in the warehouse by quarter past five on Friday afternoon or they might let you get it in by Monday breakfast time. Then they'd tighten up again and you had to have it in on Friday tea time.

When you started you used to put a tab in it, a bit of coloured thread on the side of the cloth, they used to measure how much you'd done in the warehouse.

Nowadays, they have clocks instead. We used to have great big weights hanging on the back of the looms with chains round the beam. If you had a float, a pull back, you had to fit these big weights off. It was hard for anybody old or expecting.

The weft, it weren't a bobbin like nowadays. It was wound on a tube and you used to have to put your tubes in one tin and your waste (weft) in another. Th' manager used to come round and say, "*Too much waste!*" We used to stamp on it with us feet or push it flat or put it in us pocket. I've seen some put it down the toilet. Some took it home.

I had one miscarriage before I had David. I would have liked to have two. After we'd had him we didn't want another for five years, but after that, when we didn't get one, I said once or twice to my husband, "*We'd better go and see a doctor.*" We never got round to it somehow. But I wouldn't have wanted a lot. Like mi' brother he has five, four girls, and a boy, but his wife never bothers. She just lets them do as they like. Only they do listen to their dad. He hasn't to say a lot just the tone of his voice is enough as they do as they're told.

I didn't want a lot myself because I'd seen how it was at our house with my mother. She had five. We were never short of food and we were always clean. If you only had one dress it'd be washed and ironed and ready to put on the next day, but there was never any money left for holidays or going out.

I remember we could never go on the school picnics. There were never any money.

I remember asking mi' mum for money to go and I cried when she said I couldn't go, and she cried because she hadn't money to give me. She had a right hard life did my mother, and she'd only just got to going out one night a week to th' pub wi' dad for a couple of years before she was ill and then she was ill in bed for a long time before she died. She never really had any pleasure.

Well, I wanted mine (children) to be able to do things and go places. We've always said that our David could do anything within reason. He's always been able to go on all school trips, and have a bike, and then we've taken him on holidays every year. He's never short of anything he's needed. I wouldn't have liked to have had a lot, and not been able to do things for them that they wanted.

Anyway, I will say this, we were always clean and well fed. It weren't fancy but it were enough, and there were always plenty of love at home and that's what matters.

I have no ambition to go anywhere or do anything different than what we do now.

I'm quite happy to stay and keep going to work at Felton's till I'm sixty. It's good money you know. Last week I got £48 with overtime. I paid £11/5 tax, £1/7 for pension scheme and 71p. for mi stamp. The week before I paid over £12 tax.

The pension scheme's a waste of money. Your relatives don't get it if you die, and if you don't get all your stamps you don't get a full pension when you retire and them as don't work are just as well off, because they get it off th' social security!

They even get their rent paid, so you're no better off having worked really, only that you have the satisfaction of knowing you've earned it yourself but half of ' em don't bother about that.

On a good week, I can make £35 without overtime, that is if nothing goes wrong. Things can easily go wrong like a loom being stopped, waiting for th' fitter or tackler for quite some time. That puts your efficiency down.

£35 isn't bad, considering I haven't got brains! Anyway I wouldn't like responsibility.

Like if you have a shop of your own, you have to work

when you're not selling and you are tired. You can't go on your holidays. You can stop off from weaving.

I had a friend who had a green grocer's up Laneshaw road. She used to have to work when she was ill and then when she was in bed poorly her husband had to run it for her, but he couldn't manage like she did.

He did n't know the prices, and when customers came in, they didn't chat with him and tell him all their troubles like they did with her. They were just in and out.

In th' mill you can stop off and things don't come to a standstill. I wouldn't like a job with a lot of responsibility. I couldn't stand it. If I had a hat shop and people came trying them all on and didn't buy them, I would feel like throwing people out!

Mi brother's boy goes to Hamford College. He's sitting some exams next month to go on to a college in London. He's working at Allan's supermarket at nights filling up the shelves. He gets paid five pound a week. Well it's spending money for him. They need money at that age. He's nineteen.

Mi brother got a rise at work but they stopped it in income tax. He said, they even took that £5 into account.

Mi brother Ralph, he was only about six when mi mother was ill. Every time mi mother went into hospital he went to stay wi' mi mother's sister, mi aunty May in Cheshire.

Mi mother died when he were twelve. An' he stopped going to school then. We didn't know because he used to set off after breakfast and didn't come home till tea time. Mi dad did all ways with him after we found out, but he wouldn't go. As soon as he were sixteen, like I said before, he went i' the navy. He did right well there. He won prizes and that.

But he couldn't settle when he came out.

He went in one job after another.

He's in an office now at Fishtons' at Rawtensall.

He could have done something with his life but he never did.

It was a shame really. Now he's trying to make his son do what he didn't do. I say it's not right doing that. I don't think you can learn by somebody else's mistakes. You've got to learn from your own.

Mi brother was brainy and he's just thrown his life away. George his son was brainy too. Well he went to th' grammar school. His four girls aren't brainy but they're not numb either really. But they didn't go to th' High school. It would have broken his dad's heart if George hadn't gone to th' Grammar school. He didn't have to sit for his eleven plus though. They'd abolished it where they were living then.

I'm not like that with our David. When he took his eleven plus I just said to him, "*Do your best. If you pass well and good if not, it's alright you can only try.*"

Our Ralph complained about George last week. He said he's not swotting hard enough for his exams. George says if he fails he'll go to teacher training college. His dad says he'll just have wasted two years if he does that. He's got quite a few 0 levels. He'll be nineteen in September. He's a right mod youth, long hair, one ear ring. I said to our David, "*If you come in here with all that in, you can go right through that door*".

He has a coat with all fur round it round the bottom and round the neck. David gave him a lift with it on the other day. He said he wouldn't give him a lift if it were raining as it smells when its wet!

I saw a lad last week with an ear ring in and a blond streak. My son wears "with it" clothes, -right wide pants and shoes with platform soles, but he doesn't bleach his hair or wear earrings. I think that's going too far.

Even children these days they like wearing fancy clothes. I've been a bit worried this last few months about our David. He says he wants to join th' Merchant Navy. Mi husband says ler 'im go if he's done his time (in his apprenticeship). But I don't' think he'll like it. He's a home bird.

But what can you do. Then it's no good we can't stop him.

He's old enough to do what he wants. We can't tell him not to go. Anyway, the Merchant Navy's not like the Royal Navy, he won't need to stop too long. He says he wants to get away and travel. His dad's never stopped him doing anything. They've always got on right well and been good friends. He could discuss anything with his dad.

I always told him if anything happened or he got in trouble, he should always discuss it with us straight away, - discuss his problems with us.

He's a grand lad really.

I have several relatives who 've gone abroad. I have two cousins in Australia. I wouldn't mind visiting them. I haven't seen them, or been in touch with them for years. They're older than me. They're twins. Their mother died when they were born. Their mother was mi mother's sister. Their dad gave one of 'em away to another family to look after. He looked after the other. He had first one woman and then another living with him. It was only when he heard his brother had gone to Australia he followed him there and joined him.

I wouldn't like to go abroad to stop really. I've no spirit of adventure. I've no ambition to do anything different.

I'm quite happy as I am.

Edna Sidebottom

Edna is a sprightly platinum blond winder of sixty-four.
She is immaculately made up at work with eye-shadow, lipstick
and nail varnish.

She looks considerably younger than her age.

She wears bright, youthful, sleekly fitting clothes, mini
length skirts and high heeled sandals. She often works overtime. They say at work she's a money grabber, but she denies it.

Mi mother worked i' mill in Blackburn. She had a lot of
children, I used to go watching her at work when I was eight
or nine and I had a liking to go in. There wasn't much work in
the mills in those days after the war.
I went working in a mill at Beachgate when I was twelve. I got
no money for six weeks at first when I was learning. Then I
was tenting at 10/6 a week. Then, eventually, I got 2, 3, then 4
looms.
I was working there when a strike came and I had to go and
get work in another place. Mi mother married twice. Mi dad
died in the war. She had four with him, and then four with her
second husband, and he had three. So that made eleven of us
altogether.
I couldn't get a job when the strike was on and went to work
in a waste shop just across the road.
I got married to a man in the regular army and had a baby. I
didn't work for five years then. But I had trouble. He left me
after he'd found a girl in a town where he was staying.
I went back home to mi' mother and went back to work to
bring mi son up. Then I met Joe. He used to work in a mill at
Beachgate, so I'd known him when I was a girl. He'd just got
back from the army, - North Africa.

We got married and I had a daughter when I was 39, and a son when I was 41. Soon after our Ronny was born, Joe's mother was sick. She came to live with us for four years and I looked after her. Ronny was just ready to start school when she died.

I got work at Lane shed, winding. I worked there for eighteen months and then it closed down. Then I went to another mill. The manager there had been to night school with my husband. I worked there for eight years and then it closed down, and I went to Rollings' Electrical firm.

I'd only been there three months when they declared three hundred redundant. I saw a notice at Browns' and went there, and I've been there ever since, except for the two years that I nursed my mother when she was ill, till she died.

I didn't retire at sixty because, I can't get a pension. I stopped work to nurse his mother and my mother.

My husband Joe has always been a very careful man. He didn't get chance to go to the grammar school, his dad died early and he had to go to work for the sake of his mother. But he was very hard working and went to night school. He studied everything he could, so that the children would have the best opportunity. He's in the office of Saunders mill now on the sales side.

When the children went to the High school and Grammar school we thought it was a wonderful chance. Our Ann tried very hard at school. She was determined to get somewhere. She always wanted to be able to travel and do things.

Our Ronny went in Grade C when he went to the Grammar school and later on, he moved up into the advanced class where they take their exams early. It's because of them that we keep battling on.

I've been putting overtime in, and Joe's been working weekends. They both went to university. They got grants but it cost a lot in extra things, fares and books and clothes. I didn't want them to go short.

Our Anne did right well. She went on to do a teaching diploma and then an M.A. She's working in Birmingham now but our Ronny has come home. He doesn't know what he wants to do, and it's difficult for him to get a job round here with his qualifications. He's got a temporary job i' th' mill at th' moment. His dad's right worried. But I think, he'll get something suitable in the end.

Work? I enjoyed it. It doesn't matter what they ask me to do. I never worry about it. I think I'm very lucky to be working at 64.

I'm very well paid. Compared to our youth it's a gold- mine. It hasn't done me any harm.

This year, I worked overtime quite a bit to save up for our Anne's wedding. She had a lovely wedding in Birmingham. Jim, that's her husband, - his father's a West Indian cricketer, a lovely man. Jim he's an accountant, got a really good job, he has.

It was a big reception- champagne flowing till morning. I wanted her to have the best. She's worked hard has our Anne.

Last year, it cost me quite a bit in fares for Ronny, He didn't like the flat he was staying in with three other boys at Sheffield University, and he liked to come home for the weekend every fortnight, so I used to save his fare for him and food and things for him to take back.

He's been a good lad. He's always tried his best , and he likes a bit of home comfort.

He got tired of staying in them flats.

Our Joe said I'm spoiling him though.

Nellie in a slim blond in her late thirties. She lives in a
neat little semi-detached house with her husband and two
daughters. They are planning to buy a bigger house just up the
road.

When I left school if you'd have rowed us all up, thirty-three
of us about twenty-nine went in the mill at that time in 1951.
But if you go into a school today, you won't find any girls
going in or just an odd one.
There's been so many mills closed down. My mother worked
there at the same mill I'm at now. When I were at school, I
used to go in to her, and I were more or less – well from being
about thirteen I suppose – I decided I were going in as well.
And that's how I came to go in really. I never took any
interest in anything else - only that, and I like weaving.
I've always liked it. Mind I've left once or twice. I've had one
or two different jobs in between but I've gone back
eventually. I went in a fruit shop serving an' I had to leave
there when I was expecting. And then later I went in a
supermarket as a cashier, check out operator. I liked that. I
liked both those jobs.
Then I went home-help for twelve months. Now that was for
the hours. You couldn't get the hours then really at the mill
that you can now. The hours were half eight till four and they
were right grand hours.
 I'd worked right through on shift work from the time I were
married till I were six months and a week pregnant – the time
when you come on that pension.
Now five weeks after Jane were born I went down to see the
manager for evening work.
Just at that time, there were plenty of weavers and he

didn't want any so I went to an electrical works on evening work, when Jane was five weeks old. But I only lasted six weeks I didn't like it. You see I don't seem to fit into....I didn't fit in there at that firm. My husband looked after Jane then.

Then I held out as long as I could and then I went back to see the manager again, and I ended up going back on full time. It had to be really full time. I wanted to go back because I wanted the money again. So the next door neighbour nursed her from being eight months old. And then again we used to work it in with shifts. I were on shift work then. So after a bit my husband got up and I had to leave again. I think she'd only be about three or four when I left.

Then I saw the fruit shop job advertised. It were only part time so I went after that and I got it, and I left. Now that were only three days a week and my sister looked after her those days when I worked. I've always had her. I don't think I'd have been able to work really but for my sister...the hours I've done. We never lived far away from each other. When she lived down Treefair avenue, I used to take them down and pick them up and then of course his mother used to fill in if our Anne couldn't do it. My sister really has her hands full she works hard, she does...too much I think.. I tell her so. But I do too much myself.

Our Anne didn't work then and she used to look after my two. They were at school and I finished at four so it was alright. I were at th' fruit shop when I had my second little girl. She wasn't so old and started looking in the paper again for work and saw this check out operator at a supermarket. This were only three days a week. I didn't work full-time or anything. The wage weren't enough. Like they had us on the busiest periods – Thursday, Friday, Saturday. It were hectic – Saturday especially. And the wage – it were only four pound. Then I met one of the weavers from the shed and she said well, why don't you come back to the shed. There's a lot doing part-time. I didn't have any sort of connection with it after I'd left, I met one or two of the women, "*When are you coming back*", and all that.

But in my mind I were never going back, and when this woman mentioned this part time I told Anne and she said she wouldn't mind doing a bit, because hers had gone to school. She decided she wanted to go back to work so we decided between us that we could make a shift up, sort of between us. So that's how I came to go back (to the mill). I went down to see Jack the manager and he said right away, "*Yes*"

Sheila were only little then.

We did have to both keep separate hours.

There's time now when we're both there together. But then all hers were at school. Then we did have to keep separate. Anne looked after the baby and Marion were at school and I only did part-time.

But you keep going on more, and I ended up on this ten to six work. It's all right I'm used to it. Our Anne does earlys one week. It's too much but he can't fit anything else in yet. And then she does evenings this week. Then she's started going two mornings extra. That's how you start. You do a few more hours and then... I don't know why, I can't sit back. Take me, I mean the management at work wouldn't bother. We went back on part-time, nobody's pressed us into doing any more. It's just what you do yourself really. My mother was like that. It was always a case of if you're not well go. If you're not better come home. Only once you get there you don't come home. Really I've gone to work when I shouldn't do, - when I've been ill or one of the children sick. Anne does it. We do it yet. Really it's something in you that you can't....you feel guilty when you don't go to work.

I can be really poorly and all the time I'm thinking this time, this time, I'd be doing that.

Daft really it's just how you are. We both must be born workers because as I say, we could have things easier and we don't. They get their family and they think that's it. I can't be like that.

But I don't know whether I'd rather be working there now. I think there's a lot of other opportunities I'd like to follow up.

But I don't know. I think it has a hold on you somehow because it's what I think I do best or whether I'm so … I like the atmosphere and the people. You know there isn't that atmosphere in other jobs that I've been in. I think it's a smashing atmosphere. It's like a little world of its own. I think in a mill,- mind you, I've never worked in any other mills... I don't know what it's like elsewhere.

But I've always got on well with everyone that's come to work there. And more or less I'm one of the old ones now. I feel old an' all, to a lot of these young ones. I don't know what the atmosphere is, but there is one. Then again there's the money of course. You can't get the money in other jobs that you can get there.

But it's hard work. I come home at night and I think, well, really you could do with something easier. It's just making that move, and I never seem to do it.

Now my sister, she's not really what you call a born weaver. She didn't go into weaving really because she wanted to. She were a better scholar than I were at school and she could have had a better... she could have done better as far as jobs go, but I don't know.

She's a good weaver. She does her work, but she hasn't got the feeling about it that I have.

She tells me... she says that herself, and I think really when her children do get older I think she'll be making a move. That's if she finds what she wants.

I don't know if she knows what she does want.

Now I'd like to go on school dinners really but the money isn't... You start to think well.... If you're going to put yourself out to work you think you might as well earn what you can, as play about for... you know.

Then you see another thing - shift work...

I don't think that's a good thing for women. Although a lot
of women like it. But I never did, and as I am now I have the
hours to suit me. I work ten till six, then you see I'm here to
get them off to school.

My husband's on two shifts now. When he's on two to ten,
Anne has them after school and the other week he's in by five.
It's complicated but it works in. It's a bit of a struggle some
weeks, - say my husband has to swop shifts or if they put
him on day work or change him about. But I've managed up
till now.

He's supposed to do six to two, and two to ten, but if work's
slack... like this week they've put him on half past seven to
five, now the children don't land home here till about twenty
to five. (They're twelve and seven). So they're only on their
own about twenty minutes before Bill gets in. I leave a key.
The little one could go to my sister's but for twenty minutes
it's neither here nor there. If it were longer I wouldn't leave
them. Ten to six- the hours are longer really. I could work
part-time . I'm right silly really, and yet I never do. I'm half on
one shift and half on the other - making two halves spare.

As I say I could have things a lot easier but I don't. I don't
know... I could sit back and just do part-time and really have
the ease of it, and yet I keep giving myself more hours. I
suppose you're thinking how much you could make, and what
you could do. It's all money when it boils down, you know.

I try to save a bit I suppose more than anything.

My husband worked at a rubber factory on regular nights.

He didn't like that.

It were murder really... night work.

He didn't like that so he left and had to take a drop in wages,
so I thought well, if I work ten till six full-time it'll make up
really for what he's losing. I couldn't make a do on one wage.
I couldn't stay at home. I don't know how anybody can

manage on one wage these days.

My husband is one who leaves everything to me. He never says, "*Don't do it*".

Now if I say, "*I'm tired*," he says right away, "Well don't do it". But he's not the type that's held me back. Like, if I came home and said I was working on Saturday, he wouldn't bother. He wouldn't say, ''*Now what are you doing that for?*" He just doesn't say anything really. He always says, "*It's up to you*", really.

 Luckily this isn't a big house to follow, so when I get them off to school in a morning I wash up- dust round every morning and I vacuum. I set the fire and leave it tidy generally.

 Now weekends you see... you're at it all weekends. I have to do the bedrooms, washing, ironing. I don't like to do anything through the week. So Sunday morning I wash and Sunday night I iron. Through the week I just keep on top of it really. And meals, sometimes we prepare them the night before and he comes home at five o'clock, so he more or less gets the tea cracking. Sometimes the children wait for me, sometimes he's fed them, and we have ours together. It all depends what they're having. It works out alright. In the morning I just have that bit of time to sort myself out. Sometimes if we wants something (washing) I just rinse it through and peg it out.

Oh I have done early shift but oh I don't like it. I don't know how they do it, because you've no time in a morning and when you come home you're too tired to do anything. Well, I am.

It shatters me it does, that shift. I do it very rarely. It'll be for my own convenience if I do it, because I don't like it. No it's no good, and the back shift I don't like it. In fact I don't like shift work.

I wouldn't go back on shift work. I wouldn't disrupt my family that much to take shift work on.

I always think well, I'm here at night for the children and I think that's one of the most important things. You're out at night. You don't see them. They're in bed, and then in the morning they're off to school. I don't think - I'd have to finish if it was a case of shifts.

Like if they get a lot of workers, they might want to get rid of those on part-time. But they don't seem to be getting many new workers. The young ones don't seem to be coming in. I don't know why, but they're not, they're not in any of the mills in Burnley.

In fact I were talking to one of the old tacklers, he's retired now, and his brother's a manager up at Loughcome and he said he's running it with part-timers and you know at our place, a lot of them women are over sixty. I don't know how they do it and they're even doing over-time at sixty. I mean they're in at six o'clock in the morning till...I mean they're going back to my grandma's days six i' th' morning till six o' th' night.

They make me feel lazy sometimes. I don't know, but I don't come with th' tale. Well I've nothing to bother about. I've no family, but still that's no reason, I mean its hard work in there. You're not in there to....You're at it from going in to coming out.

I mean you go for a quick smoke but you don't feel like you've had a break. You never get what I call a break.

It's a lot harder weaving the automatic looms than the Lancashire looms.

I was on the Lancashire looms four years. When I went at first it was all Lancashire looms and it were a lot easier. You hadn't that going...you must walk miles now in a day...in a week you must. But automatic weaving's better but I used to like Lancashire looms. You hadn't that urgency in your work. When you're on production in there, and you know you've got to make your percent you're on a rush and its non-stop. You might just stop and have a bit of a talk not for a long, because if you look round you'll be mucked up. You don't get much chance - you've to sort of talk and work. You've to do it automatically and it sort of comes natural after a bit. But I've noticed these younger ones - seventeens - they won't. ...They're not interested in learning to weave. They'll come in for battery filling because the money's there. They wouldn't get that money. They're just not interested. I think they're getting that much money that they're not bothered. Just doing the battery filling's nothing...you just keep going but there's no responsibility – nothing really to learn about it. Just sticking these cops in. They don't want to learn that all. I don't know what they are going to do in about ten years time when all these old ones have gone. I don't know whether I'll stop there...sometimes as I'm getting' older.... When I were younger on twenty-seven looms it didn't seem as hard to me as it does now. And I think to myself sometimes am I getting older or is this work getting rougher? I don't know.

We had thirty- six looms at one time and yet it's as hard now to me to run twenty-eight as it was to run thirty-six. I don't know whether...I have visions of myself getting something easier, when the children get to work, 'cos my husband says, "*You'll want something on mornings then?*". I mean like I've said I even think now about getting a fresh job but I never get one.

I want school meals really because of the hours. There's plenty of easier jobs I could do, and my sister could do but it's the money...you see the money, because it's certainly not the work. As I say going to work to me isn't like going to work really. It's like going home really, because I know everybody, and everybody knows me, and as I say there's that atmosphere. It doesn't do for everybody, and the noise I think it puts a lot of young ones off.

They come in, they only last a day...it's alright for us...the noise, we never think about it. These young ones coming in...I think with going in to my mother from being young it never bothered me, I used to think it were noisy but I used to like it.

But I do think it's the noise that puts a lot of them new ones off and old ones that come to try that have never done it before. But it's something I just never think about really.

Jane is a good help now she's twelve.

If they weren't sort of good children...if they were rough, I don't think I could leave them. It makes a big difference. Jane is very good. She brings her little sister home from school. And if she has to stay for choir she won't stay if she knows she's to bring her home. Alternate weeks she doesn't have to bring her home. Sometimes, they get taken to school. It's alright really. They haven't to trail to school and come home a lot on their own. It works out that there's a car available one week or another (either their dad's or their uncle's) so it's not so bad, and then they're good. I know they can be left for that twenty minutes or half an hour and they'll be alright.

I have a lot of friends at work. I don't see them outside – odd ones. But everybody mixes. You only get an odd one that doesn't mix. But there's a lot of cattiness. I think you always get that where there's women.

Everybody knows what's going on.

You think you've a secret but somebody 'll know about it. It's marvelous how things get around the shed, and it's amazing that they enlarge by the time they get back to you.

I don't know what it is about the atmosphere. Nobody bothers you so long as you are doing your work, and there's nobody over the top of you all the time.

My sister hates to say she's a weaver if she goes out for a dance with the girls. I never bother.

I said to her, "*Oh, well, tell them your something else then if you don't like.*"

I think she has a thing about it. I don't know why she went in the mill. She's not cut out for it. Now Ann is one that never mixes much. I mix with everybody. She doesn't. That must be the difference. She finds it a bit of a bind, a lot more than me.

But I don't think weavers have a bad image now. They used to have. There's a lot you wouldn't think were weavers I know when we're going home we look like God knows what...we try to look tidy but you can't... you're full of dawn and your feet get filthy.

They used to have an image but I don't think they do now. A lot of weavers...the old timers still wear clogs in the shed. Some young ones wear a modern type...coloured. They get them in the market hall. An old chap makes them in coloured leather to order.

I know when any of these bosses are walking round the mill, the managing directors...all dressed up that's when you feel down really.

Some of these young ones come in now and they look like they're going to a fancy dress parade...all their hair up, and make up, and high heels and that.

These eighteens and seventeens - they come in their right thick soles and they work in them all day. I don't know how they do it. Like, I wear high heels but I don't wear them clod hoppers. Weavers did have an image when I went in at first...you know, white fents and all that, and their clogs...but there isn't that now. It used to be smashing at work...the atmosphere I think when a lot of this older end have gone there won't be a lot of this atmosphere about it, because all these younger ones that's coming they don't mix like the older ones do. They keep to themselves in groups.

It used to be smashing for parties in the shed. We used to dress up and tie the tacklers to the poles, just for fun, and go round with mistletoe and that.

But since the automatics took over you just haven't the time. You'd have to stop your looms to do it. We did used to stop them then but we'd only four.

We used to have parties at us looms and the managing director 'd come round. Mr. Tramson always came round at Christmas...All the looms were stopped and we had parties in groups. And we had drinks and he'd come round and have a drink with everybody. But then when automatics came that were all stopped.

We asked for a party this last Christmas but we didn't get allowed one.

One or two of 'em still...but you don't enjoy it when you've to keep your looms going. All that's had to go. We asked to stop for half an hour but they said if we did we'd have to make it up and then everybody were mad because the bosses and the foreman had a party in the office. Well I mean that were maddening were that!

So some of 'em said well we're having one next year then. Whether they will or nor I don't know. We had a bit of a do on our side of the shed but it had to be in the quiet.

We used to have a lot of fun, but all that's gone. In fact, a lot of the old traditions in weaving are dying out. We used to have ware-house lads and every Christmas and Fair he'd come round collecting for himself and all the weavers used to pay for him bringing your...for being the sort of running boy between the weaver and the warehouse manager and that - you know.

That were a tradition but you see all that's stopped now. They used to come round and get a collection and we used to have a sports club and netball team. All that's gone.

I were captain of the netball team when I went at first and I used to collect the sports money. I think it were three pence a week.

We had a works' trip.

They got a table-tennis table out of the sports money.

They used to have parties and concerts for the weavers' children.

The weavers used to give a concert. One or two would sing

and one 'd play the piano. We used to go when we were
little. We used to get toffee and a pie. It were something for
the children. We still have collections for weddings and
anybody retiring but I think now it's just a matter of work.
You go in there to work and that's it.

I organize parties at Christmas. I go round to everybody. Do
you want a trifle. "*Are you joining*?" I ask them all. In my bit of
time before I get there, I go to the shop a few days before. I
order what we want and then on the day we're having it, I
dash to the shop, take everything in and give it all out before
we set on. You know it's all got to be done more or less in
your own time. Whereas on Lancashire looms we used to
have it brought in. You can't have it brought in now.

Everybody were a lot freer on Lancashire looms. I don't know
why. Perhaps it's because there were such a lot of weavers.
It's a big shed. There must have been loads of people.
Whereas now there isn't. It were a longer day then. But I
don't think they were as tired as they are on shift work.

These aren't going in the shed (the children). I wouldn't put
them in.

The only reason they'd go in th' shed would be if they
couldn't get a job and it were just a put on till they found
something else. There's a lot doing that in the mill now. You
see we're getting a lot there now, school-leavers- lads- and
as soon as they're eighteen and can earn more money they
leave. They don't want to go into it for life. It's just sort of a
pastime. They're not going to make a career out of it.

In fact I think they'll run short of tacklers as well as weavers
in time because there's no boys at all coming in the mill. They
don't stay long enough to find out whether they're going to
like it or not.

And I think it's the same in all the mills. They're all
struggling. I can't myself think that it'll die out altogether-
and then again in years to come I don't know...when there's
nobody.

I mean I have a lot of friends who left school and worked in the mill with me up to having their children and then left and they haven't gone back to it. There's a load of them really. They just have no intention of going back in.

They're doing other work. One's gone to the hospital. She's working with young babies and another one is a nursery teacher. She's as old as me and she's two children and she worked in the shed for years. I talked to her and she said she wouldn't come back and she said to me, "*Are you still there?*" Funny how they think you should be doing something different - better really.

My sister said she wouldn't put any of her girls in. I think we must have been put into it really though I never had any other ideas. My sister had but she were never able to carry it out. I mean in them days, everybody were hard up. And that were where the money were. Though it doesn't seem a lot now. But it were a lot really. It were mi mother really. If she hadn't been a weaver I doubt whether I would have gone into it. I don't suppose I would.

It got to a point where after automation came in you lost that sense of freedom, because you daren't move away from your looms or you'd have somebody onto you right away and everything you did had to be sneaky.

We still did it. Nobody were going to change over night. But you don't go away to have a smoke or owt like that if you don't think you can do. When I go out, I'm organized before I go. But it's a case of dash there, dash back and then start again.

We're supposed to get a ten minute break, but we never do. We can't. If you stopped for ten minutes it would take you an hour to pick up again. You go in and you never stop.

It's a long time really and you only get half an hour break to eat. You can brew up but it's all doing while you're working. I brew up but you've to look to see if you've time to brew up before you do. Then it's a matter of when you get to drink it.

You can't just sit quiet for five minutes in fact I carry the pot from loom to loom or else you wouldn't get it.

You wonder why you do it really. It's what you're used to I suppose.

There's no young people coming into weaving now. They must have over six people in that shed, key workers over sixty and I mean they can't go on for ever. To look at them they look as if they could go on for ever, 'cos I mean they've always been there.

They were there when I went. They think they're never going to finish. And they do overtime which is daft and I can't understand why they're doing it. Because they can get their pension and part-time and they'd be no worse off and yet they are doing all that work.

In fact, one of them said to me the other day, she's sixty-two, she said, "*We're keeping this place going with overtime.*" I said, "*Well it's not doing you any good*". But it's a waste of time talking to any of them 'cos they won't listen. But as I say I don't know what they're going to do.

Today there was a lot off. That's another thing that's come into fashion - Friday off. Once of a day everybody turned up for work whatever day it were. But nowadays - and the foreman are looking to see who's come and who hasn't. These young ones they don't come in on a Friday - these seventeens - these battery fillers.

They have the day off and they get away with it. I suppose it's because they'd be stuck if they had to sack 'em.

Now they're setting a lot of Pakistani men on, because they can't get the whites, our men - whatever you like to call them- to do the work. And I mean its work that has to be done. They don't mix at all. There's one next to me a boy, he's one you know. Now he's really nice and very friendly. I could have a laugh with him and that. We look at one another sometimes and make a face at each other because we're getting nowhere (with the work I mean - all in a mess).

I've tried to talk to him but he can't understand me and I can hardly understand him and I think well it's time to stand there and explain how you want to. You just haven't time. You shout and meemo with the noise. But you couldn't meemo to him. He wouldn't understand.

All the old tacklers who were there when I started, well they've all retired now and there's only about four tacklers who are in their thirties - none younger than them. I don't know what they're going to do. Later on - I don't know...

It used to be everything weaving, but it's like the pit. - That's another thing that's dying out. But as I say, weaving comes second (nature) to me. It's just nothing to me. Whereas I could go into another job and I'd probably find it hard work.

It is hard work. But it isn't hard work when you've done it as long as I have. You do it automatically and you don't realize it's hard work.

Whereas these new ones come in. They think it's hard. They'll probably never make it.

I find I have to concentrate a lot. Once your concentration goes you can go hay wire on your looms. You've just got to have that concentration and then you're alright, but if you're not going to put your mind on them and get organized, you've had it.

And then speed – I mean a lot of it. – Like, going on to four looms, you're just stood there. Well now it's all speed.

Before on the Lanc's looms you were just in the middle and they were there round you. Well now I mean a lot of the women they just haven't got the speed. Or they don't want to put enough into it to get any speed.

A lot of these older weavers are on part-time- but to do it a full shift....These that are over sixty on a full shift they're not weaving. I don't think they could do it.

They're weavers' helps. But I don't compare weaver helping to weaving, because you haven't got that urgency in your work. They work and they work hard but they're not

running their looms and trying to make their money.
They're on a standing wage and it's a case of how fast they want to go and how slow they want to go really. They can only do one job at a time. There are none on full-time weaving over sixty.

It's a case of trying to do all to please you now. Well we've never been used to that. Now they're stuck (for weavers) they will listen to you, where they wouldn't listen to you once of a day.

They're very good about children. They don't expect you to put yourself out. If you go and say I can't come in tomorrow or I can only do so many hours – same as me when they're on school holidays -I have to come off it. It's too much. I try to be at home part of the day.

My sister's always been in the mill but she's not the type to be in. She doesn't mix. Nobody ever knows what she's doing, whereas everybody knows what I'm doing! I'm in wi' everybody. Cheese and chalk we are.

I know all the goings on in that place. She never knows anything. It's funny. One thing- if you mention something to somebody and you say, "*Don't say it*". It will be round that shed like a fire! It's always been like that. I don't know why. It's terrible really. When you think about it everybody knows everybody else's business. I try not to tell 'em too much of mine. I don't think anybody really knows much about me. Some of them, they tell everything -silly. I always try to sort of keep in good books wi' everybody.

I never sort of...I don't put one against another. I say, "*Eeh, what a shame!*" or "*Oh! Did they?*" But I never fall out wi' anybody. They do. But I don't.

Then that were another thing on the Lancashire looms - th' Tacklers. They were...they tried to make you frightened so you wouldn't put so much work on for 'em. Well when the new manager came he stopped all that. That were a relief were that. 'Cos really some of 'em used to put on you, you know and then there used to be a lot of...

The weavers 'd just dress up any day. It didn't have to be Christmas, when we were on Lanc's looms and they'd come round- Friday afternoon - especially when the week were finishing. Sometimes they'd bring something to put on or they'd put their fent round their shoulders and just walk about in between the alleys and everybody 'd have a bit of a laugh. But I mean when we went on automatics all that were finished. It took a lot out of the tradition really.

We used to have a Christmas dinner in the canteen but then it stopped. I don't know whether the cook got fed up. This year when they said we couldn't have a party, I went in at half past nine. They'd just come in for their breakfast and everybody on one shift were sat in paper hats on. Mind you this were their party at half past nine i' th' mornin'. One o' th' weavers were on th' piano playing Christmas carols, and everybody were singing and eating their breakfast.

And I though well blinkin' heck - because they couldn't have half an hour they took their half an hour in their break for their party. And that buzzer went at two minutes to ten and everybody had to get up and go.

There were no lingerin'. But they were just sort of havin' their own little bit in their own little way. That piano's been there years an' nobody plays it.

Things are changing week by week now.

They're running it upside down actually.

Really you never know where...! Go all over the shed and apart from me, ...there's a few others that can just be put on any job.

That's why I'm spare.

I don't go on what I call my set till 2 o'clock this week, so I'm spare. This week, I've been on three different sets in the morning, day by day. There's odd ones do.

But it's a matter of sorting them out day by day on what labour they've got.

It never used to be like that. They're like taking one from here and putting them on there and they're short in every department. It must be really hard to run.

That's why we can't understand why they don't go on day work and then they'd have both shift workers.

But then they say there's a lot would leave, as they are doing these few hours at night and that.

If they went on day work they still might not have enough to do full day work, 'cos half of 'em are over sixty and on the pension. So I don't suppose they'd be better off.

Somebody came up with the idea of three shifts. But I don't know where they could get enough for night work as well, when they can't keep two shifts going.

There's one weaver there, she'll never leave. She's been there since before me. She's never have any children and she's never had a different job.

And really, she's in a position not to have to work shift work and yet she's worked shift work ever since she went on it.

She moans about it all the time and I say, ''*Well why don't you get day work?*''

Because things aren't good between her and her husband because she's on shift-work and yet she still does it.

But she's not without money and she moans about earlies and she says she's going mad because she's going nights.

She could get out o' th' mill and get a day work job for less money and still be happy.

''*No.*'' She'll say, "*I'm waiting for me redundancy.*"

I said, "you'll never get that. You'll be here till you're sixty."
I said, "I wouldn't wait for that."

She makes it an excuse.

I think she thinks she's part and parcel of th' mill and that it couldn't carry on without her.

These are the old stock.

I mean they won't have women like that in a few years because they wouldn't do shift work, if they didn't need to.

Susan Ball

Susan lives in a pleasant semi-detached, stone house on a quiet cul de sac with gardens front and back. Her flower beds are well cared for, and her rooms tastefully furnished in seventies, contemporary style, with white tiled floors and white wall cupboards. She is a pretty young woman of thirty three with a mass of auburn curls and large almond shaped, green eyes. Her slender fingers are beautifully smooth and topped with long, frosted pink nails. She has a slim, graceful figure in trousers and sweater and a quiet, demure manner. While she was talking her four children, aged nine to thirteen were coming in and out of the garden, as it was a school holiday.

I went straight to the mill from school at fifteen. I just followed my mother and sister in really, it was accepted then. It's not now. Children don't follow their parents now.
My mother had never done anything else.
It was alright but I'd only been there about twelve months when they started to change over from Lancashire looms to automatics and I went battery filling.
I found the new looms, the automatics were interesting. I wasn't on the old fashioned looms very long.
It has changed a lot.
There aren't as many people at work now. You can be on your own really, especially in the evening — apart from your battery filler.
At about sixteen, I started courting.

The money was good so we were saving up, and I never fancied sewing or factory work sitting down at an assembly line.

I liked the independence of weaving. You sort of earn your own money and if you have a bad week, you're to blame, nobody else. My first baby was born when I was twenty. So I didn't have long. I kept working till the time you get that money - maternity benefit.

My friends have worked as well really. Some didn't want to go back but they found they couldn't make a do, and they had to go back. I went back when our Brian was six months old...just evenings, and then I came to be expecting again, so I only did about six months and then I went back again after Joan was born. And then the same thing happened again. I'd only been back about six months and I was expecting again. So I'd to finish again so I called it a day, then I didn't go back after the second one.

We moved house. Then we were expecting the third one. My husband was thinking of changing his job then. But he didn't do. He was in the foundry he was thinking about going in the police force.

Then when he found I was expecting, he couldn't take a drop in wages with me losing my part time job and we'd just taken a house, a bigger one, because the one we had at first was too small after we got a boy and a girl

And I was expecting twins and it had no bathroom so we had to move really.

And then I had Peter and Jennifer. I didn't start work till they went to school at five. In between, I used to nurse my sister's children so that she could go to work. And then, it was a matter of....we were pretty hard up at the time with just one wage. So it was a matter of what I could fit in.

My husband worked shifts so I couldn't work all day.

So I went back to the shed and they sorted it out for me and my sister to work together and each do half a shift.

And David's mother looked after the children and then I looked after my sister's little girl when she was at work, and we managed like that.

We both went back part-time.

I had my sister's child, she had my bigger ones after school alternate weeks. And his mother helped. We've been carrying on like that since then, but they are all at school now so it's easier.

As the children have got older, we have both more hours in. But they still let us have hours that suit us.

My sister and me have always lived near each other. My sister has done other jobs but she didn't like them. I worked in a cafe two nights a week, Friday and Saturday for about eighteen months, when the children were little. But my husband didn't like it.

I did it just to get out of the house. It was all children all the time. I used to enjoy that, but my husband stopped me going because he said it was late night custom that was going in.

I usually walked to work when I'm on evening shift though its quite a way to walk, but my husband brings me back in the car unless I get a lift with somebody.

I do a full shift now when I'm on earlies - 6. a.m till 6.p.m. but I do a few mornings as well. 10 to 2, so it adds up to a twenty- eight hours a week.

Sometimes, I do Saturday mornings as well.

With these arrangements that I have, I'm always there when the children are going to school or when they come home.

My husband helps me a lot really.

When I'm on earlies, he'll do the ironing and things like that. When I come home at two o'clock, he'll have vacuumed and dusted.

I never let him do the washing though, 'cos he ruins too many things.

I let him do the ironing 'cos I hate it.

Thursdays he does the hall and stairs and landing for me and the girls do the bedrooms.

The oldest girl, she does my bedroom and the youngest girl does the bathroom, and the boys' room. I pay them to do that because there'd be too much falling out.

I don't want to force them into doing housework really, but it's a big help to me, and they're very keen to earn something. They'd do it for the day to come round for their job.

I pay them fifty pence each. I don't have to do the bedrooms. I've been upstairs today and I've changed all the beds. I see that they're changed every week.

They don't do that.

They've only been doing it about six weeks.

They're keeping their rooms a lot tidier because they know that they have to clean up now.

But if they ever feel fed up, I won't force them to do it.

I let the oldest girl do it, and then Jennifer came and said it wasn't fair that she couldn't do it.

She wanted a job too.

So that's how it came about really.

I don't think it does them any harm.

Brian, whose thirteen, baby sits for a friend of mine on Saturday nights.

Saturdays we do the bulk of the shopping.

My husband goes with me, and I get some things in the week on the way home from work.

I try to get enough stuff in to see me through from Saturday to Tuesday. I get all my vegetables delivered every Thursday. I have a grocery delivery as well once a week.

Sometimes, I don't know whether I'm coming or going.

During the school holidays, I found it a bit hard work.

Because, they're sort of under your feet and falling out and you're tired.

But some times, I like to go to work out of the way and then my husband shares with them.

Right from being babies, he's changed them and fed them
We've worked together.
I wouldn't like to do it all on my own. Once they get older
they get easier in some respects but harder in others...easier
that you're not having to fasten coats and wash and dress
them...harder because they have more to say for themselves
and minds of their own.
They try to tell you what they're going to do and what they
want to do, and all their school activities. You have to be
careful, because if you go to one then you have to go to
another one. You have to make sure you're fair.
 I went to a school concert last week, Jennifer was in.
Now I'll have to be sure to go to anything the others do at
school.
They have netball teams and choir and they're coming home at
 all times.
That's harder because when they were babies, you knew
where they were.
Brian's thirteen. He wants to be an electrical engineer.
My husband's going to a parents' meeting on Thursday to
discuss his next year's subjects. They're dropping subjects
now and keeping those on that they will need for the careers.
He's in the Grammar School stream. He's very happy he likes
school.
I think they show a lot more interest in children at school than
they showed when I was there.
In my day at the Secondary Modern School, I went to, they
didn't show any interest in you, when I look back. You don't
realize it at the time.
I was very happy at school I had a lot of friends. I passed the
eleven plus but my mother wouldn't let me go, which when I
look back, I think is wrong. If you went to a Secondary
Modem School there was no urgency. You didn't see any
interest somehow. My mother, she never really encouraged
me to do anything.

We never had a lot when we were little. She had always had a struggle and I think she was glad to see us start work. We had hard times. She died at fifty, just when she was getting on her feet a bit.

Jennifer talks about being a teacher. She's in the Grammar School stream. She's doing well. They all are, I know this, I wouldn't let them go in the mill.

My husband says, he wouldn't let the boys go in the foundry. Whatever they want to do I'll encourage them. I won't just sit back.

With their homework, they've passed me.

Brian's maths is just French to me. I used to help with their maths when they were at primary school but I can't now. Sometimes, I help them with their English.

They're really doing it on their own. I can't be much help to them. It doesn't seem to worry them though doesn't school. They like it.

It makes a big difference financially, me going to work.

If I didn't work, we couldn't have a car. That would have to be crossed off the list, and once you've had one, you don't like to be without. I pay for that...the petrol and everything. The color television too, I pay for that out of my wage. And then I put some towards the food.

I have so much for housekeeping off my husband, but I don't have to keep to it knowing that I'm earning. Because I'm never right good at keeping to it anyway.

If I hadn't got it, I wouldn't know where to turn on just a fixed income.

I feel sorry for those who have to.

Really, it was just as bad when I had my children little. I mean I had to go out to work eventually but its hard work going through it.

Shoes I find an item...I buy them. I try to save up for Christmas and I try to save up when Christmas is over for the summer holidays.

We go to various places....the kids like Butlins. We're going caravanning in Devon this year. We never go abroad.
Perhaps that'll come when the children are bigger, but I haven't got a longing to go.
I find food quite an item for six of us. They've all got good appetites.
I have one friend who didn't work. She put up with the hardship of going without things rather than go to work. She wouldn't work full-time.
She thought she should be there morning and night. But when the children don't suffer, - they're better off with the extra money that's coming in.
That friend got a job finally on school dinners.
She's happy with that because she gets home for holidays.
I have another friend who works part-time. Most of my friends work part- time.
Evenings is alright regularly but your husband tends to get fed up.
My husband's back shift isn't like our back shift.
He can come home when he's finished about seven or eight.
I can't rely on him being home at half-past seven or eight but he usually is. He starts on hour earlier than I do. He starts at one, I start at two. They should finish at nine. If he's home at eight, he's only gained an hour during the shift.
When the children were little, my dad used to take them out in the pram for a bit up the park.
That used to give me a break. It was a help really. It gave a bit of time to myself.
You've got to keep routine to be able to manage with several children and working.
You've got to wash on a certain day and iron the next.
This week I usually wash on Sundays and Wednesday.....but I'm washing on Tuesday because I'm working on Wednesday morning this week.

You find you've a routine. I don't think I could get through otherwise. I don't know how I'd go on if he wouldn't wash up or anything like that. I've never come across it. I don't know what I'd do.

A friend of mine, - she said her husband won't lift a finger, I don't know how I'd go on with one like that.

My husband's mother was sick when he was young, so he used to have to do it for her.

Whether that helped him to get used to doing housework, I don't know.

Sometimes, I say to my oldest boy, "*You won't know how to do nothing when you get married. Come on change yourself and try to be tidy.*" I don't want him to grow up to be hopeless.

But I think my husband does do his share.

My sister's husband's not the same really but he will make tea.

But Joe he'd wash if I wanted him to.

If I said, "*Will you just do that?*" he'd do it.

But the only time he ever washed was when I was in bed having our Brian and everything came out pink and all the woolens were shrunk.

He likes an hour out though.

He's not all family man.

He likes an hour away from us and a pint.

We go out on a Saturday and he goes to see his mother on Sunday night.

I usually stay at home and rest on Sunday.

I'm tired when

I've been working all week.

That's the day when I relax more.

At night, I like to watch television.

On Saturday night, we go to his works club.

We enjoy it.

There's dancing and there's a turn on.

Last Saturday night, we went on what we called a pub crawl. We started up at the Black Swan and we finished up down at the Pig and Whistle. There were eight of us. We always go together. We're saving up at the moment to go out to dinner together - thirty pence per couple per week. We're saving that up till we have enough. After we spend that we'll save up for another meal out around Christmas time. We go with my sister and her husband and two more couples.

We take the children out for a ride sometimes on Sundays -- to Skipton Baths or something like that. But the older children don't need us as much as they did for outings. Like today, one has gone to the baths and another has gone fishing. I like to spend a bit of time in the garden. I did quite a bit yesterday.

There aren't many young ones coming in the mill these days. There's such a lot of jobs now for women. Like my sister in law - she works at an electrical engineering place. She's sat down all the time and there's music. They don't just want to come in the mill these days with all that noise.

They're on good money but it's like ...our place - you've got to be there a bit and get speed up to earn good money. At our place you can't expect to earn good money right at the beginning. You've got to train and put your mind to it. It takes a while to get up your speed. They're impatient nowadays. They don't stay. Some do but not a lot really. Weaving I've never done anything else. I don't find it hard unless it's weaving really bad.

I think the only way you can attract more young girls is by getting them to come early ... straight from school. They have an out -dated idea of weaving.

They think it is old fashioned. I think when I look back I was the last of an era, - when you followed your mother. Most mothers don't want their girls to work as hard as they've worked if they've worked and I certainly don't want my girls to. When you come home from a full day's shift you're really tired. But it's not till you come home that you realize how tired you really are.

And yet I like the people I work with at the mill--- there's some characters around. There's plenty going on and you find out about other people.

You don't go in and not speak. You find out about people about their lives and they find out about yours. I've got a few good friends there--- older women really.

I always say I'll get out of the mill but I don't. Whether I will or not I don't know. I've often fancied shop work, but then again I don't fancy ... after I've earned good money ...working longer hours for less money and then I got offered this job at a firm ... packing goods ... sitting down all day. But when I thought about the hours 7.30 – 4.30 every day. I couldn't manage it. You don't want to be doing housework every night when you come in.

So I think if they keep giving women hours to suit themselves that's one way they'll........ It's not as strict now as it used to be in the shed . At one time, you hadn't to say this. You hadn't to do that, and if you were late so many times, you were out. That were it. But now it's a lot better.

They expect you to work...fair enough...while you are there, but you feel more like you're needed now than you ever did before. I don't know whether it's just with the shortage of weavers. They weren't bothered before, whether you worked or not.

I know one thing if I hadn't have worked in th' mill, I wouldn't have got a lot of things I've got now....better standard of living......

If I were sacked...if I were forced out of it, I wouldn't go back in another mill. I'd look round for something else. If the mill ever closes down, I won't go elsewhere. But whether I'll be happy in other work, I don't know, you see. I might want to go weaving again!

When I go out... I go on a trip twice a year...ten girls of us go. When I go dancing, they say, "What's your job?" And they look at you daft when you say you're a weaver!

They don't think there's any left. They wonder **if** you've got your clogs in your bag or something like that.

But the majority of people away from Lancashire, I don't think they realize there are any left.

When we go on this trip, the ten of us, we usually go to a club at Blackpool or something like that. I've got to know the others from one friend. I went to school with her.

We have trips from work as well. I've been on one of them but the next one coincided with our trip and I had to choose. I couldn't go to both.

I think they think if you're a weaver that you're thick. That's my impression that I've got over the years, but it's just not true. It's not everybody that can weave or run a lot of looms. It's not everybody that's successful in weaving today. There's a lot more can't manage than can manage. I think it's becoming more specialized

The type of girls...women… that people think are weavers....they won't do at all, because it's going to get more and more automated as its goes along.

Some people's work is their life but mine isn't. It's something I have to do but as soon as I walk out of the mill, I forget about my work. I forget about it, till I go again.

Some people bring their work home with them and their work becomes their life, but my home is my life and **if** the mill closed down, I wouldn't be that upset really. There are so many other things you can turn to.

But when you get married and you have children coming along its surprising how you just sort of don't say, "*Oh well, I'm going to do this and I'm not going to do that*".

You must put them first and then see what you can do yourself and how you can work things out, and hope that one day, you might be able to do some of the things you want.

I always had a fancy for nursing myself. But I can't stand the sight of blood so where do you go! If the children have a cut I daren't look.

But I wouldn't mind the nursing side of it- looking after old people, I'd like that or anybody in bed that doesn't need operations.

I did apply for a job that was advertised .....receptionist at the hospital. I found I could work in the hours with what I have to do. I wrote off for it, but I knew as soon as I posted it, (I knew) I didn't stand a chance. You had to put down what you'd done.

I put that I'd worked at Spindles all them years and been a regular worker which counts a lot. But I just knew as soon as they saw you'd been a weaver for all them years, they just wouldn't even ask you for an interview. I think that's a bit unfair really.

They looked for qualifications that aren't necessary...just because they went to High School or something like that.
I think jobs like what.....it was only receptionist....they just need somebody nice and pleasant. It was only where people come in. It weren't typing or shorthand or anything. I just fancied it.

I wrote off for another...receptionist in a solicitor's office...no typing. But you just don't get interviewed. You only get as far a writing a letter.

I think they should interview everybody who applies for such things.

They'll say, "*Which school did you go to?*"

And just because you didn't go to the High School you won't get it.

The only thing is to go back to school and get some '0' levels, but I don't think I could fit it in at the moment and work as well.

Then of course, those jobs wouldn't have such big wages, but you get to feeling like that, when you want....

Reg and Doreen

Reg and Doreen own a five roomed terrace house which they have decorated. Reg is thin and dark, Doreen plump and fair. They have one child. A rumor that they have a coloured television suspended on a chain from the ceiling proves not to be true.

Doreen . He was already working in cotton when I met him, and when he went working at this place, in Grovenham. I went there working with him. I always seemed to stick with it then. Once you've been in it. When I first left school I were in an office. I were like a telephonist, - an office worker. And I did do hair-dressing for about three years and then I gave it up. That's about all.
I've been in cotton ever since. I went in that through Reg. I don't think I'd have gone in otherwise.
We met in Manchester. Reg and his friend had a night club. I went there for a job. That's how I came to meet him. I didn't actually meet him in cotton. I went as a barmaid to the club.
My mother had a pub.
We just saw this advert' in the paper and she phoned up and I went to see about it and I met Reg there. That's how we met.
Reg. - I'm from India and I came here and worked in restaurants. I worked as a waiter.
A friend of mine worked in restaurants and then in textiles, and then I started in calico printers in Manchester. Then they closed down and I moved. I was weaver then. I worked in jacquards for three or four years and then we came to Grovenham. I was weaving all sorts. I wanted to learn how to build jacquard machines.

Then we came to Burnley to see about getting a shop and a house, but I couldn't get one. I wanted a clothing shop, shop premises and living accommodation. I couldn't get one at the price. I was lucky I got tackling after I came to Burnley two years ago.

Doreen. We decided to sell the house in Grovenham and we were going to buy a shop, and we got a bit fiddled over the price, so we got another house didn't we, and then the house we got she wanted half the land back did this woman that was selling it. So we decided to move to Burnley, but it was a spur of the moment thing, really.

But we like don't we? We got the house before the job.

Reg. Before we moved to Burnley, I came about a job.

I looked in the papers, and I got my job weaving without troubles, I was weaving and moved in straight.

Doreen . I moved in just after Reg and I got weaving didn't I? I had been cone winding at first and then I changed to weaving and moved around a bit.

Reg. I applied for tacklin'. The Manager Mr. Cross, he helped me, also some friends of mine. They helped and I was very lucky to get one coloured person in the tacklin' business.

I am nearly the first in the whole country I think, so I was very lucky. I was tried for long time to get it. I was keen of it.

If I do the job I want to be. I was so lucky. I was very thrilled of it.

And I have thanks for the manager, I am alright now.

I am nearly fourteen years in Britain now.

I came with a friend he works in textiles now too.

Doreen . You stowed away in a ship didn't you? It's a long story.

Reg. A friend of mine works in the Merchant Navy. He's been here once. He ran away.

He stayed here about two years.

He was in some fights.

He got deported.

Then I was still in school. I went to get out and have a change. I went from my home village to the town to live with my uncles and grandfather.

I was looking for a job. I liked to travel mostly. I apply for shipping companies. I couldn't get it because it was full. No vacancy at all.

Then he said he tried it. He tried first time to go. Then he gave me ideas. Do you want to go to England?

So I said I wanted to go to England, - to see and go back. Then alright, I'll take risk and both of us...because he had friends on the ship. They help some...gave us some meals. We take trouble one month... We didn't have enough food one month proper for three times, and I think Suez Canal, Aiden, Port Said, - no food at all, except water.

And I say when we came. ..shirt, trousers, jackets, that's all, nothing else. We don't carry anything, - even a single penny! He had friends here. He had a friend in Altringham and we went there.

Then he gave us the address of the Taj Mahal in Manchester and we went there. I worked with them for about two to three months and then I started to go in Textiles. Now the work is alright for me. I like it. Everything is all coming new - new parts, new machines. There are all no troubles this time. The new machines are coming these days. The other tacklers are alright with me.

I am alright with them. Fair enough...! won't bother them, as long as I've got my own job to do that's it.

Doreen. I like the weaving more than winding. I've been weaving about a year haven't I?

I asked if I could go weaving and they said I could. I'm right settled. I like it.

I've got six extra looms.

They're short of labour, so they asked us if we'd take six extra looms and do our own helper weaving, so we get a few pounds on top of our wages.

I'm right suited, - both wise and...of course it's convenient because we work on opposite shifts, so it's convenient for the little girl, because we take her back if she's off school. When one's going on, the other takes her home. So I'm settled. We go in the canteen with her. If I'm on lates, Reg takes her home. There's a few do that now, and take them to work.

She's six now. She goes to the school just up the street. We used to manage like this when she was a baby. We've always been on opposite shifts. At Grovenham, she used to go to Mrs. Tylor didn't she? She was a old age pensioner, and sometimes she'd have her or we'd swop over. Otherwise, it's worked out well hasn't it? She'd only be about six months or something like that. We've been doing it ever since. In fact I think she's been better. She doesn't follow me round so much. Before, if I went to the toilet she were behind me and cried after me. I think it's made her more independent since I've been on shifts. She once stuck more to me didn't she? Well she's more with her dad now, when I'm at work. Now she's with both of us. It works well.

Mind you, you don't see much of each other. There's changing of the guards like at two o'clock. But we wait up for each other don't we? We don't go out anywhere at all. I like reading, knitting.

I don't go out anywhere do I really? I have been to Bingo about twice since I've been to Burnley. I've no other sporting hobbies, except that I'm supposed to be in the football team and one of the girls suggested there should be a girls (team), and I'm supposed to play next Sunday. I'm a bit dubious about it, because Reg's been coming home with bruises when he's played.

Reg.   I played. I play football, table tennis. I play games for half an hour even at work for my break period for relaxation for half an hour. I forget my work for half an hour. It's a break. I play even darts anything, for half an hour.

Whenever I work I want my half an hour relaxation that's the main thing. Now I play five a side

Doreen. You've played twice now haven't you?

Reg. Yes, Saturday yesterday, I played straight from work for one hour.

Doreen. I'm a bit dubious now. They seem to be limping from one thing to another.

Reg. This fireplace? Oh yes I like do it yourself. I did all the kitchen too. I made all these shelves.

D oreen. And I do the wall papering. He does the building. He's done most of it haven't you ? The sink unit he put in and one or two other things. He's handy with his hands aren't you?

Reg. I use my brains. I try my best. Home? I have not been back before. I have a step-mother. My father is still alive, I want to go and see him before he dies. He's over seventy. I'd like to see him. I have a sister. She is married. I have a younger brother, step brother. Only me can support my father. He can rely on me. We send money.

Doreen. We both work and send money.

Reg. How much we can spare. He don't ask. He don't force me to. As much as I can spare I send. He understands. He can write English and typing. Now he has no job. He has not got a license for his business - no money..nothing at all...I have to support. Now I want to go and see him once before he pass. I've got few relatives from uncle's side and father's side and now I'd like to see them too, most of them have travelled. I will probably go for six or seven weeks. I don't know many of my country men from round here

Doreen. He does not mix much with his own.

Reg. I live on my own. I don't interfere with nobody. I have one or two friends...I don't bother. I have friends I worked with but I don't go to them. They all live in Manchester mostly. I don't go much, I like home, stay home...come home from work.

Doreen. I'd like to go home with him but really we can't afford. We'd need about a thousand pounds.

We would have liked to go into business really.

Reg . I like to have my own business...night club. Two years ago three of us had one. It closed down.

Then I went to work at Grovenham. I went into partnership. I worked both ways (textiles and pub).

I worked part-time as waiter in cafes as well as weaving. At one time, I was sick...I got no proper time to have meals or sleep.

Doreen. We work overtime now. I worked two extra mornings this week.

He works Saturday morning. You get used to it, it's surprising. I never used to like getting up at half past four in the morning.

We have mini-buses to take us from the end of the street now and it brings us home at ten o'clock after the back shift.

The only thing now is that there's nothing going on...there's no social things going on there.

That's about the only thing. I'd like to see more of these.. things going on.

One of the cotton factories I worked at before, they used to do a raffle every week...what they made out of it were used for people's children to have a party.

It breaks up the monotony, because I think in th' cotton factory it's monotonous... a lot of the jobs and such things take your mind...I don't mean take your mind off your work...but it gives you something else..you know.

I like my work.

I'm not dissatisfied now. I'm happy as I am.

I get down on my looms and that's it. In fact last week one tackler said...he were only joking... *"Those looms have done that many picks this last year you've driven 'em so hard. It's a good job you don't drive your husband like that."*

..Jokes like that.

I like it not only for the money. I like to get a good percent so I'm makin' 'em run. You have to do, don' you? You don't get paid for doing nothing.

We are going to go down to my mum for the holidays. Not actually to stay at mi' mums' - her accommodation is not so big, but my sister's just been down there.

They have a caravan and she's booking me a chalet. So we're hoping to go aren't we? Well we've never been on holiday before have we? There's always something cropped up and we've been for days. You know.

In fact I've never been on holiday in my life really ... only for days. So were hoping to go this year aren't we? All being well.

Reg . My friend, when I have time off, he always says come and help me in his café

Doreen . We rest at weekends.

We take off in the car sometimes don't we. Reg's passed his test haven't you? Well, were quite happy aren't we?

Well I am, aren't you? I mean there are things you come up against, like people saying you are married to such and such and I used to get right upset. And I used to right feel it.

Now I don't bother as long as it's not my own as are saying something.

But you know we'll be sitting round at work and then somebody 'll say something, and then they feel embarrassed when they see me sitting there.

You've to laugh it off.

It used to get me down at one time an' we get on with everybody around.

We've always got on with them haven't you?

So otherwise, I don't know.

We pull together and that's it.

Reg . The thing is we are trying now we want to get a semi detached house in a better district and then I want a little business of my own.

4. NEWCOMERS : THE YOUNGER GENERATION

In this chapter are vignettes of workers, observed from the shed floor itself by work- mates. The first six brief sketches of new workers are of women and their relationships with their husbands.

They stress the responsibilities of married life and the attempts of some women to find release from them through taking lovers or seeking separation and divorce. The impressions of the twelve teenage lovers who follow incidentally give an indication of the effects of eros and the contraceptive pill in the factory.

Some immigrants and their contacts with locals are portrayed, as well as several unusual characters and relationships between three parents and children.

These scraps of conversations, confessions and gossip depict a cross section of new workers and give some insights into the kinds of things said at work - in the weaving shed, in the winding room and the canteen; the talk which gives the mill that atmosphere already referred to and those glimpses into other people's lives, which add interest to the working day.

Many of the situations and individuals depicted portray relationships and activities which provide stories for gossip.

They include divorce; wives who have lovers; men who depend upon the wages of their women folk; youths and girls who sleep out at night and sleep at the factory in the day time and couples who live together without benefit of clergy, as well as unmarried girls with children and promiscuous teenagers.

The impression is given that discipline in the home as well as at work is nothing like it used to be. Young workers and women on big wages can afford to rent their own accommodation, away from the interference of parents or husbands and so think they can live with whom and how they please.

Gone is the respect for the older generation, for the opposite sex and for authority figures at work.

Freedom and pleasure - not work - appear to be the guiding principles of the lives of many of the young and not so young, who are determined to have a "good time", while the money and opportunity lasts.

Some older women break their domestic ties and try to enjoy financial and sexual liberation enjoyed by the younger generation, with varying degrees of suffering and success. A number of the new workers appear unhappy and ill-adjusted to their domestic situations.

Change and attempts at change are accompanied by signs of anxiety, dissatisfaction and sometimes distress. The impression is given of widespread lack of commitment to or interest in the work and a high turnover of workers, who come and go, as domestic situations change and new opportunities offer themselves. Commitment at home or at work is now more unusual.

Wives and Husbands

The Runaway Wife

A small, frail-looking woman in her forties begins work,
"I have to work", she says, "I've left my husband. I've promised
myself ever since the children were little that I'd leave him when
they got married.
I've changed my name to Jones so that he won't be able to find me,
I've got a part furnished house on May street.
I'll admit I'm a bit scared at night as I've only a bed, a table, and a
chair, and the houses on either side are empty, as the street is due to
be demolished, and only a few of the houses are occupied.
Every noise sounds louder during the night, with the house being
almost empty. She's charging me £3/10 a week but hasn't given me
a rent book yet.
I've kept a key from home. I'm going to go back when my
husband's at work and get some of my clothes. I'm smart when I've
dressed up you know. I've two fur coats."
The next day.- I've told that big women, "You're not going to boss
me you old cow!
The following week. - I've bumped into my husband down town.
He did look ill.
He said he wants me to go back and he's been worrying about me.
I've told him I'll think about it.
What would you do? Just have a look at this love letter he's sent
me with a little drawing of him and me.
That's him with the cloth cap on.

On the Pill

A young woman called Saidie starts work. She married
young and has four children. She now feels determined to
have a good time at the expense of her husband and children.

Her neighbours tell of seeing different men visiting her house and the curtain being drawn to, all day and night. Her husband works on shifts.

She brags about how many men she goes out with. Her mother says,

"She is a very naughty girl since she went on the pill"

Saidie says,

"Don't believe what she says. She's only jealous. She would be doing the same thing if she got a chance."

Her younger brother, still at school, goes to baby sit for her and takes his girl friends with him as well.

Saidie accuses him of stealing from her house. One day, Saidie looks pale.

A work-mate says,

"I've never seen such carrying on in a family. They are now even saying that Saidie's mother's youngest child is Saidie's husband's. And the mother is making accusations about her husband and Saidie, who are very attached to each other."

Saidie's father becomes ill and the mother and children argue in front of him about whether she is entitled to have another man when he has died.

The mother tells him she has every intention of getting married again and the daughters tell her they will jolly well see that she doesn't and will stop at nothing to prevent her.

The Regretful Wife

A young couple start work on alternate shifts so that they can mind the children between them, and so not need to pay nursing money for someone to look after them.

He spends a lot of money on backing horses so it's doubtful whether they are any better off.

She is a hardworking girl and, when feeling upset one day at something he had done, she said, *"Many times how I wish I had never got married."*

An Uncomplaining Wife

A middle-aged, poorly-looking woman starts work.
She is a very kind person and very willing to learn. "*I have come here because it is near home and the last place I worked at was getting me down - travelling and climbing steps at work.*
Tom has bad health. He has a bad chest so he can't work."
This was her husband who enjoyed his beer and cigarettes and called for her each week on the night she took her pay packet home.
She was visiting the hospital and had been told they would send for her to go in and stay.
She said , '*They tell me I may be a while before they send. If it's so long, I'll be under the sod.*"
When she went into hospital the people who she worked with made a collection of money to buy her bedroom slippers and a night dress for her in hospital.
One of them had seen her outside in the street grasping onto a wall near her home.
She was trying to get back after going to the shop for cigarettes and newspapers for her husband!
She died soon afterwards without one word of complaint against him.

An Erring Wife

A woman of about forty years of age with two daughters in their teens begins work today, but it is obvious that she can't concentrate on listening about her new job until she has talked about herself.
She has been working at a large garage among a lot of men and has apparently become involved with one of them.
Her husband eventually finds out and there is great trouble.
He loses control and knocks her to the ground and kicks her and then throws her out of the house.

She has to find accommodation on her own and still thinks the man wants her, but gets a nasty shock when he has to face the truth that he doesn't and she realizes she has lost her husband as well. She said,

"I've some sleeping pills in the corner of the drawer and keep wondering whether to take them or not".

The Beer Maker's Wife

A young woman who starts work today hopes she doesn't get the "shakes".
After an accident at work, she had to finish her previous job she'd done for twelve years and she felt very depressed about changing her job.
She thought of putting her head in the gas oven and told her doctor.
Her doctor said, *"It's not the end of the world".* She confides that she had a bad first husband and had to leave him - with three young children and no home.
She said she married again and the second one is no better really.
He drinks three bottles of whisky every week and gallons of beer.
He has started making his own beer to cut down on the expenses she says, *"But the crazy bugger can't wait till it's ready for drinking.*
He's so keen to get at it. He gets in such a drunken stupor that I have to change his bed sheets some mornings. They are wet through."
She said, *"I found out why my first marriage didn't work. I wasn't sexy enough."*
She becomes talkative with another woman.
She confides in her that she has other men friends, including several from Pakistan.
They leave their possessions with her when they decide to go home to Pakistan for any length of time.

She has become friendly with these men while working as a bus conductress before her present job.

She begins to have time off work and makes various kinds of excuses.

Her work mates say she meets some of these men in the next town during working hours so that her husband won't find out. There is a different excuse every time she is late or fails to come to work – sick, forgotten to lock the door and had to turn back, even fell over the child's potty and hurt her leg.

Both her daughters are separated from their husbands and have returned home to her with their children.

Friends and Lovers

Living o'er th' Brush

A very impudent boy of seventeen years of age comes to work. He starts to feel his feet and leaves a good home to go and live in a flat with his girl friend.

He tells a young woman at work. *"We're living o'er t' brush, as it were, but thinking of getting married next June"*. The following week he gives up his job for no special reason, so he will have to keep his flat running from money from the dole or National Assistance, if he doesn't get a another job.

His mother starts to look ill and worried, as he is her second son to turn out this way.

Sleeping at Work

A youth starts work and six months later is found to be having a sleep upstairs behind some bales of cotton away from his work. His services are no longer required.

It is not surprising that he's tired really. He has another job waiting on in a pub' in the evenings.

He works over-time at work and earns a very good wage. Another youth is repeatedly late for work and being pulled up by the boss.

He tells the sad story of a blind father he has to dress and make breakfast for, before he comes to work.

It proves to be a lie and he is found sleeping between some pieces of cloth, only an hour after starting work again one morning. A work mate says, "He's sleeping out wi' a bird each night and he looks for a quiet place to 'ide and sleep during th' day".

Three New Girls

Three new girls start work.

Two of them are fifteen - just left school and the other one is a seventeen year old sister to one of them.

The two young ones cannot work above an hour at a time before asking if they can go to the toilet, in other words wanting a smoke.

"Can't you manage more than an hour without a fag?" asks the supervisor. But they don't seem to be able to.

One is very pale with an expressionless face and the other is very heedless when told to do anything.

When they are moved during their second week the supervisor says,

"Me and Ada have been on aspirin all week for headache through dealing with these, plus three more school leavers.

I keep taking them one at a time into the hoist (due to the noise in the shed).

It's taking me all my time to keep my hands off 'em. I swore at one and then realized she could "have me" for it.

She cheeks me back and says, "Alright then but stop shouting at me!".

The older sister at seventeen comes back to work after a shift with her baby.

A young man is familiar with her who works in another department. *"Is that your brother?"* I venture to ask.

"No", she replies and then a short pause. *"He just lives at our house"*.

Two Girl Friends

Gillian, a well-built girl of sixteen begins work.
She wants a change from her last sitting- down job.
She says she is unhappy at home .
Her father and mother are divorced and she lives with her father and step-mother in one town and her sister lives with her mother and step-father in another town.
Her step-mother seems determined to stop all her pleasures, which are games. As she said, she is not really interested in boys.
She buys a cycle at so much a week and pleads with her stepmother to let her have a dog. Eventually, she agrees, so long as she pays for its food. She won't let her go to pictures or other things, like the rest of the girls.
She becomes very friendly and obsessed with a young woman Kathleen who has come to the mill from a neighbouring town. Kathleen dresses in men's clothing and is often mistaken for a young man.
 She soon has Gillian doing whatever she wants. She seems more infatuated with her than a new boyfriend.
She goes off food and smokes a cigarette instead of having a meal at break time. They become inseparable and Gillian goes to sleep at her place at weekends.
It seems that Kathleen only has a weekend place though, which is with another friend. She sleeps rough at the Salvation Army during the week.
Kathleen was brought up by foster parents and has been knocked about. She has a little girl who is not living with her and seems to be 'in care'. She leaves work after a few weeks as she says it's not altogether the work, but she can't

concentrate due to not having proper accommodation.
She said the council had promised her something but she
would have to wait.
A few days after Kathleen finished work, Gillian finished as
well and went to join her friend.

On National Assistance

Two young brothers start work. The older one says ,
*"I'm saving up for a car. My dad and me are sharing at one.
He's not working he has a bad back. My mother cleans up at the
hospital"*.
A few weeks later, his sister Margaret starts work. She is
separated from her husband, gone back home.
Her dad is looking after her baby while she works.
 She is still only about twenty. During the first few weeks, she
complains of feeling off color. *"The doctor has changed me onto a
different pill,"* she says.
She has become attached to a sixteen year old youth working
on the upper floor.
After a few weeks they were reported to be living together in a
condemned house. They pass the baby to each other at the
end of the shift, being on opposite shifts.
He irritates the other workers by telling them that when she is
away from work she can draw as much as a wage from the
National Assistance, after telling them that she lives on her
own with a baby to keep.
Eventually, she decides to finish work and take life easy on
National Assistance.
Edith a big cheeky girl has been friendly with this Margaret
and she also decides that she will finish work and asks her if
she can come and live with her little girl. Margaret agrees, but
when Edith's mother finds out she does everything in her
power to stop her from going. The mother has enjoyed some
good wages from Edith by encouraging her to ask for
overtime work.

Edith picks up with a big unkempt Irish man, who already has a wife in Ireland and lives in with another family with the daughter, who has his baby.

Edith tells her friend, *"I'm going out with him."* The chance to live in with her friend gives her an opportunity to see more of him.

An Unwed Mother

A tall girl Irene is very difficult to train, perhaps due to the fact that her mother refuses to let her go out at night, since she had an illegitimate child.

She repeats *"My mother's scared to death of me coming home with another child".*

She arrives at work today in tears,

"I was only out of the house a quarter of an hour this morning and my wage had been pinched and both meters broken and emptied.

Do you think the boss will let me have a "sub" till next week?"

One week, she is especially 'off color', the reason being that, the wife of her child's father has started work in the same room.

"I'm going to tell her he's behind with my payments," she says to a friend.

"You'll be in trouble if you do", said the friend, as his wife didn't even know of the existence of this child.

One worker said,

"I hope she doesn't get to know or there'll be skin and hair flying".

Irene seems to have days of acute depression and feels that everyone is against her.

She sometimes decides to go home with a headache or a bad cold in the middle of her shift work.

If the foreman pays a visit to her home to find the cause of her absence, her mother makes a point to keep repeating,

"She will have to come to work. She has a child to keep,"- determined that she does not let her forget it.

Eventually, things look brighter.

She seems happier and manages to smile. *"She's courting,"* says the supervisor. *"She's found a chap at last. He used to work here."*

A Teenage Father

Michael starts work. He has a very fresh complexion and straight, sandy coloured hair.

 He has tattoos on his arms and tries hard to create an impression with the other boys. *"I'll be alright if I watch my step."*

A young girl starts work a week later and says he is her husband.

 "He is only seventeen," she says, *"and we have a baby 8 months old.*

 He had joined the army but he was recalled to marry me. I am working on the opposite shift to him so that we can look after the baby in turns.

 But he won't be seen pushing the pram to work, so I have to pay a friend £2/5 a week to mind the child 2 hours a day.

 He won't let me bring it to work.

 "Is he good with you and the child?"

 "He's alright," she says, *"but he has a terrible temper which he gets up for the least little thing."*

She soon finishes work and makes the journey to work at pay day, where he is seen to give her a 1 pound note from his wage packet.

She looks very poorly and unkempt and cold one day so I ask him why. *"She hasn't been well,"* he says.

"Has she got a mother?" I ask, feeling a bit worried about his ability to look after them.

 ''Oh yes she has a mother'', he says.

"She speaks to her but she has no time for me."

"Well! See that you look after her and the baby,' I say.

Not many months after he says, *"I'll have to be off work for a week she's expecting another nipper.*

It's supposed to have arrived by now. I don't know..I'll have to put a bomb under her."

Left Holding the Baby

A teenage girl in very mini skirts with a lot of make up on and bleached hair, and looking much older, begins work.
She said, *''I'm going up to my boy friend's mother for breakfast, we have a baby but he is going to marry me. An old lady is minding the baby for me."*
Another girl starts work who lives in the same street as the girl, *"I've learnt a lesson from seeing her"* she says.

"Her boy friend is two timing her and she has just found out. I doubt if he will ever marry her and he accuses her of neglecting the child.

She was turned out of her home when expecting the child by her father and stepmother. Now she is on her own with the baby. Her step mother left her dad. Then her dad went off with another woman. She was left alone with the baby and she's only eighteen.
(The baby was later taken to hospital sick from neglect.)

An Artist's Keeper

June lives with two young men.
One is the owner of the house and parted from his wife and the other one she lives with as man and wife.

"He is a very good artist you know and I am only working until he sells some of his pictures".

He waits for her outside the works on pay day and she admits to the friends she works with that she gives him all her pay.

"You're wrong in your head", one of them tells her. *"What do you want to do that for working to keep him, while he is*

having a lazy time?''
She is very strongly attracted to him and despite her parents
being very upset and trying to get her away from him she
thinks he is marvelous.
Her dad is so upset. He threatens if ever he sees him, he will
use a knife on him. She continues to live with him and
"keep" him and they eventually move to another town.

A Casanova

Tom comes as a labourer.
He has good wages to draw as he works a lot of overtime,
but he has a habit of calling in at the bookies and putting bets
on after drawing his wages. This week he lost 20 pounds all
in one go and vowed he would have to keep away.
Being easily led he talked about asking if he could work on a
different shift so that the bookies would be closed, as he
passed with his wages on pay day. His friends say he has a
small house of his own although he is not married and has
parents. He lives in it from time to time, when he picks up a
girl who is willing to live in it with him. If he finds that she
is lazy or only wants to live off his earnings, he slings her out
and finds another girl.
 He often gets into trouble with the police on his weekend
escapades. He tells of a great scare he had when working on a
sea wall at a holiday resort during the summer.
He enjoyed playing practical jokes on work mates but this
time, he said, *"I bit off more than I could chew and was never as
terrified in my whole life."*
A big powerfully built man was in charge as foreman, so that
he could get tough with any unruly ones. Feeling in the
mood for a joke, Tom filled the foreman's wellington boots
with cement. This was too much for the foreman. He took
one look at him and said, *"Well I daren't touch you or I'll kill
you! You're such a little puny thing, but I have a better way of
dealing with the likes of you."*

He puts him in a skip, fastened the lid on and hoisted him up for so long then lifted it and dropped it again. He continued doing this until he was just about at the far end and frightened to death, "*Now*", he says, "*that's how I deal with the likes of you!*"

On his recent holiday he was a passenger in a car with friends, he decided that the water under a bridge looked tempting for a swim. He opened the car door and jumped over the bridge straight into the water.

He said, "*I had to do a quick get away as all the fishermen were after me.*"

Sometimes, he drinks so much beer on a Sunday he is in a complete stupor, when he should be getting up for work on a Monday early.

One of his latest girl friends started out as a typist up at Newcastle and somehow drifted away from home, arrived in this town jobless and homeless and was picked up by him to live with him, and he quickly got a job for her.

She floated off again after a few weeks and he reinstated another girl.

He now accommodates several young men as lodgers, and it's common knowledge that several girls sleep there at night.

It's rumored that one 16 year old girl working in the mill stays there some nights and arrives at work, not too clean and disheveled.

Another girl arrives at work wearing a dress the length, of a blouse.

'*You're a disgrace,*" says the supervisor. *"Where's your overall to cover your bottom. Everybody's looking at you.*"

"I've forgot to put it on," she says. *"I've left it round the corner up the street where I stayed last night.*"

The fast life seems to be finally catching up with him. He starts to have a lot of time off work and finally leaves to take a summer job abroad.

He says he was getting fed up with the girls keep turning up knocking on the door!

Engaged Three Times

An attractive girl of nineteen with nice red hair started work.
She says, 'I'm not interested in boys much.
 I've been engaged three times and broken it off. The last time I
called it off the day before I should have got married. I feel like
having a good time.'
It's not many days before she sees a young man she fancies
working in the room above.
 She starts going out with him and decides to leave home
and take a flat, and another engagement is in the offing.
She makes a habit of taking money from him for new clothes.
He is good- natured and can't see that this is the reason she
goes out with him.
But one of his work mates decides to put a stop to it by
telling her that he has seen him in town with an attractive
blond, which isn't true.
But she believes it and 'finishes' with him.
The friend feels sure that he has done him a good turn
getting rid of her.
The next news about her is that she is pregnant and getting
married.
''Who is she marrying?'' one of her work mates asks.
 "She's marrying a farmer," the mother of one of her intimate
friends say.
"But it's certain he isn't the father, as she told our June it belongs
to the man upstairs!"
Her sixteen year old sister is making dates each day to meet
a boy with long frizzy hair, pale face and gold chains round
his neck, who is said to be taking drugs.
 Her supervisor feels worried for her, so she tells her older
brother to have a talk with her, to stop seeing him.
They say that she is sleeping with different boys and feels
proud to show them love bites all over her body one
morning.

Easy Game

When a new girl begins work, the boys eye her over to see whether they like her or not, or if she has already got a boy friend.

If she is one who all the boys hang around. They presume it is because she is "easy game."

One particular new girl was well known for being free and easy with the boys. As one boy put it, *"She doesn't know how to say ''No'' that one."*

The younger ones organized a trip to the seaside and there were plenty of rumors buzzing around about what went on and this girl's name was mentioned most. It turned out that she was pregnant as a result, and it was said that her parents were horrified at what had happened to their innocent daughter!

The news was that her father went round to the homes of most of the boys who had been on the trip trying to put the blame on one of them.

It was said that she was so generous with a lot of them, it would be difficult to pin the blame on any one.

The Hell's Angel and Sir Terry

Two new men start work today.

One is called Roy and feels proud to say he is a "Hells Angel." He is tall, spotty faced and ugly with long, smelly, wavy hair which needs a wash.

He wears a black leather jacket with a lot of fringe round the yoke and bottom edge.

He has flashy studs, swastikas and chains round his neck and shabby, denim trousers with frayed edges.

Despite all this the supervisor says he is very nice to talk to and much more polite than most trainees.

He is eager to please and on hearing a repeat of the talk on safety, and that long hair should be tied back, he says, *"All right I've got the message, you're hinting at me."*

"What the hell have we got now?" one of the work men says when he sees him.

"Don't be put off by his appearance", says the supervisor. *"He's really nicer than he looks".*

He doesn't stay long. He gets into trouble with the police and is kept in the police station several nights, till he can prove something about a stolen motor bike. He comes to work to explain and blames one of his gang and says, *"Just wait till I see him!"*

The second one is called Terry and is very pale and keeps nattering about all kinds of things and says he is getting a divorce from his wife. When writing his name on his locker instead of putting Terry Cousins, he writes 'Sir Terry'.

He works alright for several days and then takes a day off. On his return his instructress says,

"Where were you yesterday?"

"Mind your own bloody business", he says.

It came out in the newspaper the following day he had been at court for altering his sum of money in his bank book. It said he had only recently come out of prison. He never came back to work.

Odd Characters

A Self Confessed Witch

A haggard looking young woman with bleached, straw-like hair and heavy make starts work upstairs and looks like she has never been to bed. She talks about leading a very fast life. She shows us photos of her wedding which are unbelievable. She has changed so much in a short time from a pretty girl.

Her daughter of twelve doesn't live with her.

She gets friends to let her sleep at their houses and during break times she calls at work for money from her mother. Her mother can be seen making a fuss as though she was the best mother in the world.

"*Luv*" this and "*Luv*" that and "*Tara chuck, see you tomorrow*". She frequents one of the lowest pubs in the centre of town and brags about having a lot to drink, then stripping her blouse off and saying, *"How's that for a nice pair."*

She also claims to be a witch, but stresses not a bad witch but a good witch. She says she has a second sense and can tell what is going to happen in the future. Some of the workers are inclined to believe her. She certainly seems to have the power to tell people what is going to happen and it comes true.

She is really very kind by nature and a young man Lawrence, who works with her and gets turned out by his wife, calls at her house in the early hours, to see if she will take him in. Lawrence is married to a very bold girl, who has a twin sister and they bully people and scare them to death. He tells his work mates he has to sleep with the pair of them, as they won't be separated.

He looks very thin and neglected.

His wife eventually throws him out for good and finds another man. He wasn't actually sure that the wedding was on, until the very morning it happened. She was more interested in his pay packet than him. She took all his money and refused to keep going out to work and stayed in bed most of the day.

The Pick Up

A young girl, only seventeen is known to be able to drink as many pints of beer as any man and smokes too.
If any "new" young man comes to work, she does not wait to be noticed.

She asks them to take her out and is reported to be very fast.

A rumor goes round that she has arrived home in the early hours with clothing and stockings torn and is supposed to have been picked up by men in a car, thinking she was going to be brought home and instead they took her in another direction.

She tells some tale about escaping by going into a ladies toilet and through a small window by the back way.

On being spoken to by her parents she decides to leave home and goes to stay with a sister.

This doesn't seem very satisfactory as the sister is living with a young man lodger and the husband is the one who has had to leave home! She only stays for a few nights and then returns home.

Immigrants and Locals

An Irish Family

An Irish family arrives in town and is fortunate to get a small terrace house to rent in a rather poor area near the works. This is a good opportunity for the mother and father and two oldest girls to find work and earn some decent wages to take home.

The second daughter begins first and appears a bright girl but untidy and has a rather large supply of swear words in her vocabulary. Several times she was found to have written 'English Pigs' where she sat at work.

They seem to have moved around Ireland and lived rough, - the father fishing and poaching and shooting for food. The eldest girl, small with a very baby face and looking about thirteen, tried desperately because she was already pregnant when leaving Ireland, which was proved after she had worked for several weeks.

The father, was a keen worker, only when he felt like it and didn't like being tied down to a starting time.

Each pay day meant getting soaked in beer, spending up, and then demanding money from second girl Maureen. If she refused to give money, he gave her a punch with his feet, if she wasn't quick enough to get out of the way. Her 'spending money' she received, after giving her parents her wage, was soon spent.

Not having been used to owning an amount of money and being of a generous nature, she called on all the children in the street each week and made a little party for them of sweets, cakes etc.

They all soon got used to pay day and were waiting for her as she arrived in the street.

She was going out with a boy called Jim so she decided to buy a new white skirt to go out in. The following week she wore it for work as black as the fire back and creased up.

One morning Maureen arrived at work at 6 a.m. She looks disheveled and confided in a workmate that she had not been home since the previous day.

During the morning the women police arrive at work. She had been reported missing.

Apparently, she had been to a football match in the evening and then decided to sleep out in a field above the town. It was a very cold night. She said her boy friend had provided her with a blanket from his home.

Maureen was still only eighteen when she drifted away from work and soon had a baby and was unable to go out to work. She also drifted away from home.

Several months later, the child was found to have died, while left in the care of a woman in the same house and Maureen was out late at night.

Soon, all the family pack up working, as they can't cope with discipline and time keeping - another large family drawing Social Security benefits.

Living it Up

Several Irish girls arrive to start work, after seeing an advertisement in the newspaper. They come from quiet villages in Ireland with not much to attract the teenager. So when they arrive, they have a whale of a time with their big wage packets and new found freedom. They live together in a flat and live it up, appearing at work for 6 a.m. looking rather disheveled and tired, looking like they've only been in bed for about two hours.

The Suicide

A very thin, highly strung woman in her early twenties starts work. She is separated from her husband and living with a Pakistani man with her two children. She is a clean and exceptionally quick worker, *"I wish the bloody thing would go faster"*, she says as she pushes a trolley round distributing during her work.
They had recently moved from another town.
She repeated these words," *I've a lot on my mind"*, and several weeks later she was found, having taken her life and that of her children as well. Her mother had been caring for the children in a near by town. She only recently had sent them back to live with their mother.

Enjoying a Free Life

A Pakistani man begins work.
He likes to gain attention and doesn't appear to intend taking on a full load of work. People are very kind and helpful to him. He shows some peculiar ways. For instance, a worker close-by came to complain, looking very angry and holding her scissors halfway down the blades.

"If that so and so tries to trip me up again these points 'll go right in him and I'm dead serious".

To a very nice pleasant woman he says, *"I love you, if you come with me I give you a happy life".*

He had said that he had a wife and six children in Pakistan, but seemed to be enjoying his free life, watching and shouting at the young girls as they passed by. Eventually, leaving work after not making any progress, he meets a man from the factory in the building where you claim National Assistance.

"What are you doing here?" He says.

"It's more like - what are you doing here?" says the local man back to him!

Mrs Sen

A very haggard and tough looking girl begins work. Her eyes don't meet yours as she talks. She is only in her mid twenties but looks years older. She is very thin and round shouldered but full of energy.

She wears a wedding ring and pretends to be Mrs. Sen, but isn't really, as she is not married to the Pakistani young man she lives with and has four children.

She was proud to bring a photograph to work of him and his brother-two very handsome, smart young men.

She is very tough, as she arrives at work one day black and blue from head to foot, and tells her mates in the toilets he has thrown her down the stairs from top to bottom.

It seems she has to go out to work for a wage to keep them and he sends a lot of money home to Pakistan.

She enjoys the piles of tea cakes with meat on them that she brought to work. As she tells her friends he won't let her eat English food in the house, as she has to eat the same food as him.

Aside from all this, she is very satisfied with her way of life.

She thinks a lot about her children. But then she suspects he is going back home and taking the children with him. So after one of the many beatings she gets from him, she decides to leave him and go back to her mother's and take the children.

An Indian Lady

A young Indian lady starts work and complains about having to sweep the machines down, as she doesn't like the dust, *"I don't know how to sweep up"* she says,
"Well I'll show you" says the supervisor!
''That's the brush head and that's the handle, now get hold of it and get cracking".
She explains that she was a school teacher in India, and having several servants, she never did housework.
She came over to England with her husband, who is a doctor and three children.
Her husband has deserted her and is living in London. She was left rather helpless in a little house on her own with her four children, everything being very strange to her and she was finding it hard to cope.
It was sad to hear from her how it was affecting her children at school. They were shy and had language problems, as really they were clever and sensitive children.
She still insists that she can't sweep up and claims that she never does it at home, as her daughter does it for her.
When made to do so at work, she makes peculiar circular movements with the brush, expecting the supervisor to let her off doing the sweeping. She decides, on the invitation from her husband, who lives in a big house with another woman, to go and see if it will work out, if she lives in the same house, separately with her children. Then he would be able to help with their education and see more of them, as he is comfortably well off. She realizes it will be hard and inconvenient for her to see him with the other woman.

She is however willing to accept this for the benefit of the children. They return, as she said the children are growing up and will not accept the situation of seeing their father with another woman.

A Mixed Marriage

A bright, young, little Indian is settling down and doing well with his work. His stumbling block seems to be that he married a local girl from a disreputable kind of family. He lives in a notorious street, where there is fighting and arguing going on every day and night. She starts work on the opposite shift from him and he looks after the children when she works. Some of her relatives start work as well. There are soon arguments and a fight in the toilets between his wife and a woman who is already working there.

She starts on about these coloured foreigners living off a well known cat food and what she thinks of anyone who married one etc. The wife knows how to deal with her, as she is used to weilding her fists. It usually ends up that way when they go in the local pubs and have a few drinks. This episode gives those who witness it something to talk about. Some of them speculate how the woman's daughter will turn out. She is a very unattractive lump of a girl always complaining of headache and apparently kept under strict supervision by her mother.

An Entrepreneur

A Pakistani man walks about at work wearing his industrial gloves and avoiding work where possible.
He brags about having done some good deals buying and selling property since coming to England.
"How long have you lived here?" I ask.
Two fingers go up. *"Two"*, he says."

"I buy three houses:- one for 200 pounds - sell for 450 pounds; one for 250 pounds - sell for 350 pounds" etc.
"You will have lots of money then," I say. *"Oh No!"* he says. *"Send it home to Pakistan.*
" A tackler says to the man, *"What the hell has ta' got on thi' hair? It doesn't half stink. Tha' wants to pack that in and use Brylcream."*

A Handsome Bachelor

A very handsome, smart and obliging Pakistani young man starts work on a skilled job. He has been brought up and attended school in this country, so understands the Lancashire people very well.

He makes a very good impression and the workers appear to respect him and show an interest in him.

He isn't married, so there is curiosity about his sex life.

A tackler asks him, *"You're not married then?"*

"No."

"Well, how do you go on then, for a bit of the necessary?"

"Oh alright, he says,

"I have my girl friends, but when I decide to marry I shall marry a girl from Pakistan chosen by my family."

This seemed to satisfy the curiosity of the others interested.

"He's a bit of alright," one girl says.

"He's like Omar Sherif."

"More like th' Aga Khan, when he gets all toffed up," says another.

"He's no different than the rest", says some one else, when they heard he was meeting the notorious girl of the works after the 10 p.m. shift, with his car.

He walked away with a look of disgust, when the same chap asked him, *"Is it true th' art tailing Maureen then?"*

Ada says she reckons that's why he's left the works and she's not daft.

He couldn't face up to everybody knowing he was going out with this girl, after the false show he put on.

Parents & Children

A Generous Son

Brian a tall, fresh looking youth started work this week, a bright, hard-working lad, very friendly and generous – wanting to buy small presents for his work mates.
It was obvious to see that Brian needed someone he felt he could trust, to confide about things at home, that were worrying him.
His father and mother weren't on speaking terms, due to the fact that his dad spent every night at the pub, entertaining his friends and drinking. Naturally his wife and family of five children were the ones to suffer and Brian was eager to help his mum with money.
One could see this was on his mind a lot. It was worrying to see his face change one time as he said, "*If mi' dad ever lays a finger on mi' mum, I'll kill him*".
Brian said his elder brother left home at sixteen, due to having trouble with his dad. He is now married at 19. His sister was also made to leave home. Brian gave his mother extra money to help keep some of the younger children.
" *Mi dad doesn't sleep wi' mi' mother and I can hear 'em arguing when he comes in drunk. I have to turn my radio up to drown the noise. I had a row with him last night and nearly struck him. He accused me of telling our John and Irene to call him ''thing'' instead of dad*".
He tried to keep away from his home as much as possible so as to avoid him. It was an anxious time when he arrived at work one Monday morning and said to his foreman, "*He's dead. "He died last night.*"
"*Who's dead?*"
"*You know-THING*".
"*Do you mean your dad?*"
"*Yes*", he says...

He had returned home after an evening in the pub and died suddenly.

"I'm not bothered at all", said the boy. *"I didn't feel anything for him! I hate him."*

His sister now arrives home for some nights.

He said, *"She has about four regular boy friends 'on the go', but she means to have a good time. She keeps her flat on and comes and goes."*

He said. *"My mum tells her to marry the one with the most money."*

Mother and Daughter

A certain woman has a sick husband who is unable to work and one teenage daughter. She works very hard and is not a friendly person. Her family are always having rows and not speaking to each other. A new worker who follows her at the change of the shift begins to turn the pick clocks before the two o'clock buzzer goes.

"Leave them damned clocks alone till the buzzer goes", She says, *"You're taking my picks."*

Her daughter starts work and is more keen on catching a boy friend then doing her work. She injures her arm through not switching off a machine as shown when learning.

"I was watching George." She says, *"instead of thinking what I was doing."*

She gets a boy friend, who has his own flat and takes his breakfast to him most mornings. He gets in trouble and has a quick wedding but her mother makes a point of denying this. She buys a house and furniture for them in the same street that she lives in, and intends to dominate their lives.

She has a shock one day when she calls at the house and they have done a moonlight flit and moved away without telling her.

"She's bloody well had it with me now for all time", she says.

It is some time before she finds out where they have gone to. She sticks to her word and won't forgive her.

In the mean time, her husband dies and she soon picks up with a man who is parted from his wife and five children and they move away so that his wife can't claim from him.

The Foster Child

A young girl walks about with her head crouched down into her shoulders and causes her supervisor some trouble, by keep changing her mind when being trained.

She does very well with the job for a while and then lacks confidence and wants to come off the job onto something easier.

The supervisor says she is very mixed up and feels sympathetic towards her.

She also has a lot of days off work.

All this seems due to the fact that she had a troubled childhood.

"*My dad was a drunkard*", she said, "*and my mother had a very bad life. She had to work for money to keep us with and she was always poorly and in pain. She put off going to the doctor till we had all left school. Then it was too late to be cured. She has cancer of the womb and has now to be wheeled about in a wheel chair.*

I went to live with a foster mother who was kind to me and had two children of her own who are both backward and attend a school for backward children.

I was happy there till I started courting with Bill.

Then she turned against me and when I said I was going to get married she fell out with me and wouldn't come to our wedding.

I was 21 and he was a nice boy but she has never spoken to me since and it broke my heart.

It's the one thing I pray for that she will be friends with me again. My husband even is getting fed up with me keeping being ill and stopping off work.

I think, she didn't want me to leave her as I went everywhere with her. We were good mates."

The foster mother works here too and doesn't act like a grown woman.

When approaching some of the men she works with, she has a habit of saying, *"Come on give us a kiss"*.

One unmarried chap didn't think it was funny.

"I'll put my foot behind you in a minute if you don't go away", he said.

Eventually, the girl returns to work part-time, after having a baby and seems much happier and more settled down.

6. CONVERSATION PIECES

The pieces of conversation recorded verbatim between weavers, winders, over-lookers and cloth lookers in this last chapter, show some more of the characteristic dry humor of Lancashire workers; always ready to have a good laugh at human weaknesses and funny situations which occur at work and at home. Most of the talk is about work and workers or about women and their romantic, domestic and sexual involvements. Those set in the factory are about the proverbially obtuse behavior of loom over lookers, the activities of a new works committee and about new workers and foreigners. Pastimes discussed include slimming clubs, sauna baths and bingo; all new organizations which sprung up to cater for the leisure tastes of the new rich weavers and winders.

Finally come the exchanges about girls and women and sex, the perennial topic of interest for both the men and women at work; the subjects of gossip which circulates so rapidly round the shed and probably improve with the telling. This talk about people everyone knows is the stuff of which the characteristic atmosphere of the mill is made, which gives the workers the feeling of being on familiar home ground.

Work & Workers

Weaving,.....it's th' only trade that's always been on piece work and where men and women have always had the same wage.

....

"Are you coming on these looms?"
"I've never seen such an 'opeless tackler in all my life. I wouldn't care but he thinks he's God's Gift. He still brags about having worked in Brazil. My God I wish he'd stopped there!"

....

"Do you know what t' tackler's just done, when I was writing the loom fault down on his paper. He gave me a push with his elbow and said, "Don't write a bloody essay." But I said, "Don't you push me."

....

"What's up Marylin? What are you bending over backwards for? Have you hurt yourself?"
"No." Bob th' tackler says," I've got odd legs. So I were just lookin'."

....

I used to work with a tackler, they called him Births Marriages and Deaths. He always wanted to know t' cats tail out t' joint. Though he wasn't bad really, as long as he got to know all that was going on, especially the morbid news.

....

Did you hear about the Chinese restaurant down Stoney Street? A chap got a bone fast in his throat and had to go to hospital and when they examined the bone it was a dog's bone!

....

This tackler were going to work and he had 'is bait under his arm. He were going down the street an' he bethought hissel' as he didn't locked th' door, so he picked up this piece o' wire and went back and started poking under the door. One of his weavers went past and said, "What ever are you doing?"

He said, "I'm trying to get mi' key. I came without locking th' door. I just pulled it too and I shoved th' key through th' letter box."

....

Women are always a lot better than men at weaving aren't they? Take Tom for example. He always blamed the looms and the warps for him having such a low efficiency each week. He heard about a married woman earning 10 pounds a week more than he did, so he went up to the office to complain.

"Alright" the manager said, "You can change looms with her if you're not satisfied and see how you go on".

"Just watch! He'll say they changed all th' warps when he dropped th' efficiency by 30%."

He was mad when somebody said he needed space to see the cloth faults. "Does ta mean I'm not a good weaver?" he said.

"Wey, ya can expect it of him. Did ta hear about him puttin' silver cashews in his bike wheel when t' ball bearings got worn? A real gawp he is!"

....

We had one chap worked in t' warehouse and he were a bad one. He were off work one day and they collected for a wreath. I don't think there were really any malice in it. It were just to let 'im know how much he were disliked."

....

'There were one tackler he'd throw his shuttles right down the shed. You'd be frightened to death of going for 'em.

....

'There were one tackler. They called him Long John, as they said he could take the weights off th' backs o' th' looms to take up a smash from th' front of th' loom!

....

There was one weaver who only had one arm. He lost his

other in the first work war. He only had a stump and he used to run six looms with a tenter to help him. He used to put in his own smashes and he used to put th' shuttles under th' stump of 'is arm to hold it and he skewered th' cop on with 'is 'and. He couldn't knot th' twist but he could roll it together so that it would hold fast."

....

One morning, Amy and her mother Annie were late for work. They only lived just up th' street from th' mill and when they got to the shed door, Billy Hardcastle, the manager were standing there holding his watch, and pointing at it. He were a bad un.
"And Annie said, "Eh, that's nice, is it gold?"
Many a time they'd send you back if you were late. But she got in, her daughter and all!

....

"You want something better for your children than you've had yourself. If you've been oined (troubled) while you're working, you don't want your children to do the same."
"I don't think it's a matter of being oined. I've enjoyed shed life."

....

"The management used to be too supercilious. The only time they spoke was if you were late for work.
There's a big improvement these days as far as the management and workers are concerned. You can talk to them."

....

"When they're building new factories, they build them in such silly places. Look at that Welmersdale. It's been there years and they haven't been able to get workers and those what they've got are having disputes. They built an estate there and all, they never did fill it. Why build a new mill when there are all these mills here?"

"We used to say this in the union. Why build new mills when there are all these here and the experienced workers as well?"

....

"They've been on a standing wage of about 40 pounds for three years since we got these new looms in. It makes a difference you know. They just go and sit in the rest room when they feel like it. They nip back to see their work's all right and then they go back in the rest room. It'll be a change next week when they go on the bonus system."

....

"Oh I don't know they say they're not bothered about the bonus system so long as they got the basic standing wage."
"The management don't seem to realize if their wages depended on their efficiency they'd be turning a lot more cloth off."

....

"I call the weavers here greedy. They're money grabbers. They're always lookin' to see what sort are being put in other folks' looms and then they complain they've got better sorts than them. They're frightened their own efficiency won't be so high and that somebody 'll get a bigger wage than them."

....

"We used to be paid poundage at Williamsons, I preferred that when your wage were according to the efficiency of your set.
 I heard big Sid saying wages should be related to the efficiency of the whole shed yesterday.
But he's not good. He can't tackle. There's about fifty smashes a week on his set!"

....

"Have you ever seen them jet looms what squirt water Lil? They don't have shuttles at all and all the cloth's wet through. I'd like 'em I wouldn't mind a go. The weaver said it took her a week to learn how to thread it."

....

"I think this works' committee's a good idea. It's the first mill I've been in that's had one. I'm on it. I think it's a good thing. I wouldn't say there were a right lot of progress made. But there's things that you can't hurry and I suppose there's things going on that we don't know about but there just seems to be no progress made.
Although, we get to say what we think, and we get information that we wouldn't know. There are small things being done. There are improvements being made.
Things like more towels, better soap, and that.
But mostly, its conditions of work that come up. Things are improving slowly. They 're moving in th' right direction."

....

"Are you on th' works committee?" "No. Florry is, isn't she?"
"They bring up all sorts of points don't they?"
"Florry 'll be better than Gloria won't she. She does talk sensible doesn't she?"
"They have one spokesman for each department."
"I told Florry to go on. She's done very well though has Florry. Now Gloria were all wrong. She were asking for more work!"
"She's stupid though is Gloria. Silly!"
"Florry was a bit nervous at first but she's alright now."
"Well that's understandable because all th' men are there - th' foreman, the manager, th' managing directors! It is a bit of a facing isn't it?"
"But they used to manage without such like. Don't you think it's a waste of money?"
"They get a pound each all them on th' committee."
"What do they get it for?"
"Just sittin' there."
"All in the works time?"
"They're on a good thing aren't they?"
"Oh yes! A pound each, and tea and biscuits. That'd be why Gloria went on in th' first place."

....

"There's a lot of tripe talked about at them committee meetings. Them top bosses think that we don't know any different."

"They've all come here finding themselves top jobs and pretending they're running th' place, after th' weaving manager's done all th' donkey work."

"There's a lot more chiefs than Indians these days". It's workers we want, not people walking about in suits and ties doing nothing."

"I told one of 'em, one boss is enough."

....

"He looks sixty or more."

"I doubt if he'll make it. I saw him yesterday spend ten minutes at the back of the loom trying to tie one thread."

....

"The 6-1O pm or 5.30 to 9.30 shift? Oh, they call that the housewives' shift. They do alright at ours an' all. They've 11.50 pounds a week flat rate. They'd soon stop if they had to make it with their own efficiency.

....

"The back shift? They call it the savers' shift. Well you don't get much chance to spend money do you when you're working 2- 10p.m."

"Oh, I don't know about that. Some come in their dancing frocks — not only the young ones either, and I've known 'em bring cases of clothes to change into and their boy friends 'll be waiting outside ready to pick 'em up."

....

"Did you see in that place, they have a rest-room where you can go and sit down any time you like with tables and benches?"

"Some go and sit there all night on the back shift, when the management's not about."

"Well. It's better than our place. Some even take their

---

sandwiches in the toilets and sit on over-turned weft boxes. I don't know how they can do it. It turns mi' stomach over to think about it!"

....

"I want to speak to you....I'm givin' mi' notice in."
"Oh what for?"
"Well I'm not getting paid mi' maintenance from that bloke so they said at the Social Security that I'd have a better chance of getting' if it I weren't working, so I'm finishing work."

....

"Nellie's hurt her hand again. She's had to go to first aid. They'd be better-off paying her out."
"I don't know why they don't put her on t' stripping machine."
"If they put her on there, she'd fall **off t'** bloody thing."

....

"Are ya working Saturday morning Joan?"
"Yer bloody kiddin' not likely, I've had enough. I'll get me feet up and have a magnetic arse all weekend!"

....

"Do you see that new woman who's just started? I knew her when she was younger. My God, her husband must have knocked hell out of her. She used to go in the same pub as us with her friend and she was one o' th' smartest an' th' best looking girls i' th' town. All 'th men turned their heads to look at her. He's oined her to death and she has about five kids now and has to **work** full-time."
"She's slow, but she's willing.
She has difficulty keeping up with her work, but she never leaves it or has time to look up. She seems like she's tired all the time."

....

"Do you remember that Charlie, th' labourer in th' winding room?

Th' girls used to think he was a bit numb an' felt sorry for him. He got th' winders believin' as he had no mother and were badly done to, and they used to bring his lunches. They used to get 'im hot pies. They even helped him wi' his work w' feeling sorry for him, as he didn't look so strong. They used to fill th' hoppers wi' perns for him.
It were several weeks before they found out as he'd been climbing out over th' gate on th' back shift and goin' up to th' Mermaid for a drink.
It'd been goin' on for a while, before th' manager found 'im climbin' over th' back gate and then they realized they'd really been taken for a ride!"

....

"What do you think? I feel like going to see our local MP. That foreign woman who has just started work has been telling me she claims money to send her parents abroad and when my husband died after paying a full stamp all his life, I couldn't draw a damned thing. It makes my blood boil. What some of 'em get away with I don't know how they do it."

....

"That Italian woman knows where her bread's buttered. She just about pleases herself. She thinks she hasn't to obey any of the rules we do.
Have you seen her doing her own tackling and even grinding her own shuttles on the machine which has a notice up plainly saying NO UNAUTHORIZED PERSON TO USE THIS MACHINE. She's a clever bugger if ever there was one."
"Did you see her make a rude gesture when somebody said what would happen if the manager came by?"
"She's parted from her husband you know. Her two daughters used to live with her and as soon as one left school she told me,
"I'll put her on the pill I'm having no trouble with her."

"Aye and she has a young fellow almost half her age, who stays at her house on and off."

"No wonder her daughters got fed up with her and left home as soon as they could."

....

''Th' labourers are getting worse, and worse. One hasn't turned up after break
His girl-friend 's packed him in so he hasn't come, and th' other refuses point blank to do anything."

"Only in his own time and at his own speed."

"He's a bad one he is. It's not long ago him and his mate beat a boy up and made a right mess of him just for kicks".

....

"We can't work with this new labourer, he's sat there at th' side o'th' frames staring at us and undressing us with his eyes".

....

"There were trouble in t' winding room today. There's been such a row upstairs. That big-headed mechanic said I was paid to tell tales and that I was a big liar.
So I lost my temper and called him a big slut.
I realize I shouldn't have said it but he gets me so mad.
They all hate him up there.
He tries to boss us about.
A bit later on, he said to his mate *"Look at that big lazy buggar"*.
So I said, *"You have it coming to you"*. I went straight to tell the manager but he got to him first and told his side of the tale.
He swore he never said anything about me. But I told him.
I were that upset I wept.
The Manager had to give me a piece of fent to wipe me face with. I were in such a mess."

....

227

"Do you know what Moira said in th' office yesterday? She said that th'only thing they use their heads for in th' shed is for putting hats on!"
"It's a good thing for her they didn't hear her. She'd have been in for it."

....

"Tom upstairs is getting carried away with himself since he's been made shop steward. Mavis and Jean say they're fed up with him keep picking on them and being bossy, he's only the labourer."

....

"That lad who keeps coming around here, I think he must fancy little Jane. I wish he'd get on with his own work."

....

"Once over there were just like a madhouse weren't it in the shed? They're quite happy now the girls really.
I don't know, but I blamed a lot on Harry and Joe for letting it get out of hand.
They had that Lizzy and that tall Agnes, an' they just let 'em do what they wanted, while th' manager were off.
An' you see when they've got to that stage of they can do what they want wi' th' foreman – there's no work in 'em at all is there?
An' I think that's what they did. And you see then they both left at the finish.
Well, you see Lizzy she'd got that used to doing what she wanted that she said, *"I'm going on days. I don't like shifts"*. But the manager said, *"You 're not"*.
So she finished. You see th' foreman 'ad given ' em that much rope.
An' Agnes, she picked up wi' Gilbert and she just buggared off and went to live with him."

....

"Our Maureen - she learnt short hand and typin' when she left school and worked in an office for a couple of years. But she didn't like.
She could n't keep out o' the mill.
Well I mean the wages are so much better. She's been getting twice as much on them Unifil automatics as she did in that office. She does like."

....

"That supervisor doesn't like me because I knew her when she lived next to th' mill and wasn't in such good doings. Since she went to live higher up, out of town, she doesn't want to remember. Her mother used to go up with th' jug to th' pub every day and they were poor like us."

....

"I 'm about fed up I've been called names on this job that I'd never even heard of before!"

....

"Don't some people fuss about food? That Helen brings a piece of butter wrapped up and says "*Can I have this put on my bread as I don't like margarine?*"
"*Sorry luv,*" I said, "*but you'll have to put your own on we can't do anything special just for you.*"

....

"The foreman asked me to take some more weft out, but I told him "*You can sod off! I'm working over-time today.*"
"Yes but you're being paid for it aren't you?"

....

"Our Yvonne has to go in hospital.
She wouldn't go and see the doctor because she didn't want to miss going out at nights and having a good time and she's lost a lot of weight.
Her boyfriend Raf says it's for her own good.
He said she'll be looking for a wooden overcoat if she doesn't go!"

....

"Oh I'm having trouble with that Marlene. I'll only have to tell her that she's making a mistake an' all th' lot 'll be out. She thinks you're pickin' on her if you say she's made a mistake. But she's hopeless really!"

....

"I've seen one or two there today as 'ad left before, and then they've come back..."
"Oh aye' they go all round and then they come back when they haven't found a better shop...."

....

"I've got some lovely best steak for dinner off Jean. She brings it to work to sell. It's only 2/- a pound."
"You've got a nerve buying it off her. Her husband works at a butcher's in town. I bet he pinches it, and you'll get into trouble for receiving it, if he's found out!"

....

"That Fred, that young Pakistani lad's quite a bright lad isn't he?"
"Of course he went to school here you know."
"He's as good as any we have, but he's too big for his clogs by far. Did you see him poking the supervisor last week and telling her to get back to her job because she were talking to somebody!"

....

"There's a Pakistan chap outside. He said he should have seen you at ten o'clock but he couldn't make it."
"I know buggar all about it but I'll go and see him."
"Good morning. Take a seat."
"Good morning sir. I saw the boss yesterday and he say come back at ten, Thursday and he have job for me."
"Well, I'm the boss, I've not seen you before. And I've no job. Describe the person you saw here."
"He was a very small man."

"Well, there's plenty of small men here, but they're not the boss. I'm afraid we have no vacancy."

....

"It's a right shame about mi' granma. She were living in a street down by th' bus station and some come an' lived next door. They were alright in 't day time. All were quiet but at night it were bedlam-noises and shoutin' and going up and down stairs. And they used to take their curtains down at night to use for sheets. And there were young women calling at night.
Well she got that fed up while she moved....She went to live in the same street as our Bob. Well would you believe it. She'd only been there three months when another lot went to live next door to her and they'd a lot of carry on at night an' all."

....

"I say's to this fella, "Look at this cloth it's making broken picks", and he says, "Oh me not weave that. Me not speak to tackler".
He took a dive in front of me and before I know what I were doing he was round the back pulling all the cloth off the roller. I was dumb-founded.
"What did you do? You didn't stand for that did you?"
"Oh I was too taken aback to do anything."
"Don't they get away with it!"

....

"They're far and away past us ....all the tricks they do. We just stand looking and don't do owt about it. I used to bend over backwards to be open-minded about 'em when they came in th' mill at first. But I've learnt different since I've had a basin full I can tell you."

....

"Mr Brown, I have problem," "Yes Stan, what is the problem."

"Well, you see sir, I am very good time keeper, very good attendance. You see....."

"Yes, but what is the problem"" "I want to be over looker...."

"Just a minute Stan. In my mind you will be an over looker...."

"Yes, yes, I can be?"

"Just a minute. Keep your nose to the grindstone and make a good weaver first." "Oh, I do, I do, Mr. Brown."

"You know when you started you wanted to be weaver. You have to be battery filler, weft distributor, cloth remover, beam wheeler and loom sweeper. All these things you did before learning to weave. Now again carry on good weaving and in time I will see that you get to be an over looker."

"Oh you are very good man Mr. Brown. Allah, will be kind to you.'

....

"Did you hear Sid Beams told these Pakis they were sacked? They were off down th' road and somebody told th' manager. He said, "I'm the only chap as 'll do any sackin' round here." "Oh, I were only coddin (t e a s i n g) 'em," he said.

....

"You remember Sam that labourer who wouldn't work and kept putting his coat on and going home? Well he came to ask back yesterday but they wouldn't take him, even though they're short of workers."

"That chap's a disgrace for carrying on like that. His wife's working full- time." "She does over time an' all."

"She'll need to, married to him!"

....

"You know Mrs. Philling?"

"Her whose husband lives in th' next street?" "Yes well she had to finish work you know." "What for?"

"Well she kept taking time off. She was so keen on going to play Bingo every day. She put it before her work. She now does part-time work at two places so she has time to go every day. She spends a pound or two every day."
"Why doesn't she live with her husband?"
"She said she wants to go back to him, but she says he's a nuisance. He expects her to keep waiting on him hand and foot and she won't do!"

....

"The White Bull? - Isn't that where Jack goes in his leatherette jacket and his hair all pomaded down?"
"Yes,"
"His son's at work an' all now i' th' shed." "Is he alright?"
"He seems to be..." "Does Jack wait on?"
"When he's on a back shift, he works there in th' mornin'. He works down in th' cellar where they get th' beer from. He brings th' barrels up..."
"That 'll be hard work..."
"Well, he likes drink. He's always drinkin!"
"He smells of it."
"He says he never 'as anything to eat! He says, "I don't know what it is to have a Sunday dinner."
"He must be an alcoholic!"
"He said his wife never makes a Sunday dinner because he won't eat it. When he's on the six to two shift he goes straight down there, unless he has to do a bit of overtime and then he goes down there afterwards."
"His wife won't see much of him then will she?"

....

"Our Harry's in a bad temper. His wife's told him she's expecting again and he didn't want any more as he's just learnt to drive. And he wanted 'er to start work so that 'e could buy a car.
"Well aren't you as much to blame as 'er?" I said to 'im.

"No", he said. "She looked after that side of things and she's let me down."

"Don't you think she's done it on purpose because she doesn't want to go out to work again?"

"Oh it'll wear off when the baby's born. Then he'll forget and make a fuss of it."

....

"What art ta' carrying that load o' shopping for Fred? What's th' job then?"

"I have it to do. Th' wife's pissed of an' left me, left th' kids an' all.

....

"Did you hear why Mona's off? Her mother put a stop to her living with her friend so she won't come to work. She told her, "If I can't go out wi' chaps I'm not working."

"You know they've been re-housed in a council house up on hill top?"

"That'll be a change from Foundry Lane won't it."

"You know her mother. She's that house-proud. She thinks it's wonderful."

"Yes but she's worried she might have to leave."

"Why?"

"Well Mona keeps having different men calling and bringing her home. She said she's afraid th' neighbours 'll complain about it. They're such samples. She said you wouldn't believe. Even a midget from th' circus she brought home th' other night."

"She's chap mad."

....

"If you don't leave my husband alone, I'll cut your bleeding head off. And that applies to your sister as well."

....

"Did you hear about Eva. She's in hospital. She came home after staying several nights wi' another chap and her husband left her black and blue...marks all over her she had.

234

She says she has hemorrhage with his thumping her with his fists."

"And do you know what her mother said?
She said she couldn't do with her at her house.
She said her work came before her.
She said it was her independence that went to 'er 'ead after she started work."

"Well believe me she spends half her wage on her men friends."

"She's talking about leaving her husband and children."

"She won't get as many men wanting to go with her if she leaves her husband. They'll be afraid of getting caught!"

....

"A couple of queers they are, Agnes and Nellie, living together. They never bother wi' chaps."

"Did you hear what Bob did th' other night when he were going home? He stuck a cone wi' yarn wrapped round it through their letter box with a note pinned to it,
"How is this for size?"

"What happened?"

"Well they were that bloody mad they played hell the next day. They must have suspected it were him as did it."

....

"It must be a problem courting wi' a twin, wi' a sister as just looks th' same. I don't know how tha' can tell th' difference between 'em!"

"Ah but I can. That's Lil wi' t' biggest tits. I must admit though th' other night I made a mistake. I went in and put mi' arm round her and give her a squeeze. Well I thought it were her but it weren't. It were her sister Doris!"

"And what did she say then?" She said, "get lost!"

....

"Have you heard th' latest about Pauline? She's picked up with that new labourer. She says she's a month pregnant and she's going to ask th' manager for a lighter job."

"Him, he came here from Padiham following that little Christine who works upstairs."
"Well, you wouldn't think so. Every time him and Pauline get a chance they pop up in th' lift together."
"He's a fast worker. He's only been here a month!"
"She's never expecting. All she's expecting is a pair of kippers. She's saying it so as that lad won't leave her."

....

"I'm getting engaged Mr. Smith." "Oh, and who's the lucky girl?" "Mary Lord."
"And what does your mother say?"
"Oh she doesn't know yet."
"But don't you think it's a bit foolish to rush into getting engaged. I mean you're only just sixteen. Why not take her out for a while first?"
"Oh no. She wants a ring."

.... (A Month Later) ....

"How's the engagement then Bob?"
"Oh it's off. She didn't like the ring. She gave it me back. She said it made 'er her finger go green."
"What's on now then?"
"Oh I'm saving up for a car for mi' birthday. I'll be seventeen after Christmas.

....

"Jean 'll be late in. They've tied her to a post outside work and put a white nightdress on her and flowers on her head. She's getting married on Saturday.
She's seventeen and he's eighteen tomorrow."
"I wouldn't get married young again.
I just feel ready for marriage not at twenty-five."
"You can say that again.
I'm only twenty-eight and I've been married eleven years. We must have been daft i' th' head!"

....

"Have you heard that Jack's bothering with Enid. He's never away from her and she's been seen going to his house when his wife's not in."

"She's hard up. He borrows money from her every week. He never has any money with his drinking and backing horses." "What does his wife think about it?"

"Uh she got upset with the rumors. But she's very easy going you know. Yesterday, she sat next to me at lunch and she says do you want a stick of celery? It makes you more sexy."

....

"What will your daughter say about you getting married again Will?"

"I don't care a buggar what she says. It's nothing to do with her. I'll please miself. I know she doesn't want me. I went at Christmas but I soon came back. I know when I'm not wanted. I think her husband was frightened of getting lumbered with me.
But he doesn't need to bother.
I'll go in a home before I'll go there.
It was just little things they did. But it was enough for me to notice. I'm not so thick.

....

"Where will you live when you get married then?"

"Ethel doesn't want to give her house up and I've more sense than give mine up. I think we'll finish up a month or two at her place then a month or two at mine!"

....

"What do you think? Tom Blank who lives several doors away from me went to work on the early shift and had to go back home for something he'd forgotten, gets home and finds a Pakistani in bed with his wife. He gets hold of him and pitches him clean out of the bedroom window into the street below!"

"You should hear the sort of things they come out with these days. Our Marlene was with some of her friends the other day and another girl turned round and said,
"Are you all Lezes then?"
"No," she said, "and you shut up, you old Pros."
A mean at eleven, you wouldn't think they'd talk like that would you? It was a scream the other night, she brought some friends in the front room an' I looked in and says,
"Eh I'm not havin' any of that carry on." An' our Marlene says, "What do you mean mum? Hush! She's a girl!" I thought one of 'em were a lad. They'd all got their jeans and jumpers on and I thought one had a lad's face.
I did feel daft, so to cover up I said, "Oh yes, I can see now I look at your eyes, you're a little girl!"

....

"I called at Martha's today to see how she was, after coming out of hospital. She's going up the wall wanting to come back to work.
There's no room to swing a cat round in their house, and he's driving her mad. You know what Fred's like.
He's so faddy he keeps going over everything she does and then they start an argument.
She misses the money an' all, as she spends a lot of it on her two married daughters and the children."

....

"He only heard a fortnight since that she has no money. He thought she were rolling in it. So he's asked her twice since if she wants to call the wedding off!"
"She said she wanted a new dress last week and he said,
"Tha' can have owt tha wants if tha's brass to pay for it."
"Well he's looking quite poorly now and the wedding's supposed to be comin' off in three days time. He's been working overtime five nights a week since Christmas and redecorating all his house from top to bottom."

"Well, you can't keep carrying on like that when you 're over sixty-five can you?"
"I thought he looked quite poorly when I saw him last week."
"Perhaps the excitement's too much for him!"
"She's not making it any easier.
She said last week as she didn't like the fire place in the living room and he took it out and put it back how she said she wanted it, and then she said she wanted gas instead of electricity."
"I laughed. He said he'd been to the Co-op to get a 22 carat ring and he'd got three books of stamps and Lizzy said,
"Oh is she not using one of her old rings?"
I meant she's buried two.
She'll have a fourth yet. You mark my words."
"They're not married yet. I think myself he'll call it off at th' last minute."

....

"They're pulling all them houses down next door to you aren't they?"
"I've been ready for moving for a long time now. I'm fed up with that one next door. She has men and boys calling all the day through. One young lad didn't know where to look this morning.
I was standing on mi' front step when he came and knocked on her door. "You'll have to wait your turn love,' I said. "She's upstairs with one now." He walked away without saying a word! He looked some and sheepish I can tell you."

....

"I just ran into Ada in th' toilets. There she were all dressed up-Mascara, mini skirt an' the lot. You'd never think she's sixty."
"Aye an' she works harder than most o' th' young uns."
"She was just doing herself up to go dancing. She said she has a stout every night and goes dancing on Fridays."

"Dancing! After a week o' th' back shift I'm buggared. I can tell you. Give me bed right now. I could sleep on a clothes line."

....

"No wonder Maureen is popular with the boys. Have you heard the boys have found out that she's easy game? They tell one another and most of the young boys in the shed have had a go."

....

"Eh luv! How do you manage looking after three little children and working over-time?"
"I have a good husband.
He helps me a lot and he has even had the vasectomy operation so that I don't need to go on the pill."

....

"Have you heard about Peter in the office? He's going out with Jennifer." "I've heard she's pregnant. I don't know if it's true."
"There'll be trouble if his wife finds out though."
"I've known him for years and she always did have a lot of running after him before he married her. He was always keen on the girls and never missed any chances."

....

"I said to Marlene, "Why don't you have an abortion?" I mean this 'll be her fourth. She'll only be able to work another month or two. But she said,
"My husband wouldn't let me. He'll leave me if I do."
"Well what can you expect? She was having her third before he would marry her! He said he didn't want tying down!"
"They're all the same these young Irish lads."

....

"We've got two real ones started this week. They live in digs together. One is a real girl. He bleaches his hair, walks about like a girl."
"Aye and did you see how he brags about his friend having bought him a silver ring. He shows off with it round th' shed."

....

"Fancy Connie loosing' her widow's pension over a chap
like that." "And then he went and left her after all!"
I mean she didn't need to have married him did she!"
"And with losin' her pension, she's had to come back to
work full-time at sixty."

....

"How's little Rosemary going on?"
"With all th' married men?"
"What's she been doin' now? Is she still bothering with that
one in th' warehouse? Who's she after now?"
"Oh anybody, anybody that comes in. She's writing to a
boy. He was a nice boy. Th' manager found 'im a job.
He was home on leave.
You know Dick th' sweeper, well he was his boy home on
leave from th' army and they found him work. He's writing
to her".
"Dick's wife's come battery filling did you know?
It was their daughter that married that dirt-track rider.
Their Tom were ever such a nice lad. He doesn't want to get
himself mixed up wi' that Rosemary."

....

"What did you think about June Blog – married five days she
was?" "Did you hear about that? Did you know him?"
"Him, he's always been a very heavy drinker. There's never
much happiness when they're like that.
He just walked in and said,
"Oh I've got fed up", and went."
"She took him back the week after!"
"Well, she's wrong in her head isn't she."
"She'd had a right bad do with her nerves before with her
first husband, and then she takes this on. You wouldn't
credit it would you. She got a right jew."

"Well, her mother said to me she was right upset. Because she'd been paying her bills for ages. Backin' her out and payin' her bills. An' you can see she thought she'd got rid of it when she got married."

"Well she shouldn't have need to (pay bills). She's workin' and she should have known."

"Well, I think she's been on part-time,"

"She should still have been able to manage." "She did make a mess of it didn't she?"

"He won't do her any good will he?"

"Well, of course his father was worse than him."

"His mother and father spoilt him. I know that his mother said when he left, "Am I glad I'm getting him off my hands!" He'd been living with her till then. June were getting' him!"

"That didn't sound so good then did it, for his mother?"

"That was his own mother."

"Then it's a wonder how he'll get on with June's daughter.- That age I mean.=It's an awful big risk."

"Is she about fifteen?" ———

"Yes."

"He's a lot younger than June you know." "Is he?"

"Oh yes. He's a lot younger than June."

"Well he'd be more right for the daughter then!"

"Well, I'll say this, she's took her work on. Because I mean, you haven't to be bad minded but it's there isn't it? I wouldn't have taken it on. I'd have studied mi child more."

" 'Cos she's just at a funny age isn't she?"

"Yes she's just about fifteen or sixteen and nowadays they're really grown up at that age."

"She has made a mess of it."

"Oh yes, he's much younger than her. She's about 38 and it'll take him all his time to be thirty."

"Some can't tell when they're getting' married."

"When she's made one mistake you'd think she'd have been more careful."

"Somebody told me Gladys were badly done to at home. Did you hear about that, about her havin' to tip all her money up?"

"Its not true, according to her brother-in-law, Ron. He said she's a devil. We always used to think she were put on wi' her mother didn't we?"

"Well she used to give Ron her wage." "Perhaps it was so she wouldn't lose it."

"Can't she hold her own wage? She's about thirty-two?"
"She is like that though.

"He said it's not so. She only tells that story. He should know. He said it's her what oins her mother. So I don't know."

"I only know what Gladys 'as said."

"You can take it with a pinch of salt."

"I do know that when she worked beams to me they used to go on holidays twice a year,
Her mother and father and there were only Gladys in th' house what was in work and they used to leave Gladys at home to look after th' dog.
There was only Gladys working, because her mother and father was on th' pension.
She must have been tipping her money up."

"I wonder what they think of her going to live wi' Roger then, an' him a married man."

"Oh she's not staying at home now.
I asked her if she was going to see her mother and she said,
"Am I buggery! I'm not going to see er."

"Has, she fell out wi' her mother?"

"She's not spoken to her mother since she met him at Holmwood Club.
She used to go there with her mum and dad. That's how they went out. At this Club or something they must have dancing as well.

Gladys got up dancing wi' Roger and her mother said she hadn't to dance with him."

"What at thirty-two? Her mother told her who to dance with!"

"Oh yes, she said she hadn't to. So Gladys, when she went back to th' table Roger said her mother wouldn't talk to her because she were talkin' to Roger.

So of course when it's time to go home, instead of going home with her mother, Gladys went th' other way and went and slept wi' Roger! So she's slept wi' Roger ever since! She hasn't been home. So it's gone on now for a few weeks.

"It's a funny relationship isn't it? Where are they now?"

"Well they're at Stonegate house. I think there's only men there really."

"Yes, it's like a doss house isn't i t?"

"Yes, she's one there living with 'im."

EPILOGUE
Forgotten Voices

Since the stories and conversations recorded and presented here were collected, there have been various attempts to depict textile workers lives. A recent notable example is Hall's (2012) book, which has sought to capture forgotten voices of Britain's post - war working class. His workers' stories include textile mill girls, who risked injury in dirty, noisy weaving sheds. In his chapter 3, *Where Cotton was King,* he describes how in the mid twentieth century nowhere in Britain had a bigger concentration of industrial sites than Lancashire and how Lancashire cotton dominated world markets in textiles. He reminds us that already by the mid nineteenth century Lancashire was Britain's main textile manufacturing base. The period he focuses on is the decade of the fifties, when Britain was still a major industrial power, with nearly three quarters of its more than twenty million work force classified as manual workers in "working-class" jobs. The urban centers in which the working classes worked, such as Burnley, were centers of black smoke, belching from scores of chimneys. There was little greenery in sight and little fresh air to breath, other than during the holiday (wakes) week spent in Blackpool or other seaside spots (such as Butlin's holiday camps).

"people's memories of Lancashire in the 1950s are of

a landscape full of cotton mills, coal mines, steelworks, canals, railway lines and locomotive sheds.

In mill towns like Bolton and Burnley, mighty centres of filth and fumes, steam would be rising and smoke scowling as hundreds of chimneys belched forth day and night over the tightly packed terraces of small Victorian houses (Hall 2012:134).

Now as Hall notes the few remaining textile mills produce special products, such as flame retardant fabrics, coated textiles and breathable materials, so that by 2010 there were still around 4,300 textile workers in Lancashire and a further 600 or so working in the clothing industry, making mainly specialist work wear and uniforms. And the buildings housing these production processes are now clean and modern industrial units, not big dirty mills with belching chimneys, polluting the air. While Victorian terrace houses still remain, there are many housing estates to accommodate an increasingly immigrant population.

Textile Families

The stories told in this work do not just record daily working lives of isolated individuals, but reveal family relationships and domestic scenes, as well as relationships in the factory. There are stories of matriarchs and devoted, aspiring couples, of devoted dads and feckless parents. There are accounts of sisters, who share their child rearing, of lovers who share their partners and there are images of immigrants, who adopt local customs and learn local techniques. There are scenes depicted of the young, who revolt from the strict discipline and asceticism of earlier generations. Survival was the paramount goal in earlier times, tempered by pride and ambition and a desire of many to promote the best interests of the younger generation. A repeated theme is thwarted educational ambitions of the adult workers, who were blocked from taking up grammar and high school places by family poverty.

In contrast there are examples of sacrifice by parents to make sure selected children enjoyed the chances they never did, to go to (private) secondary school and college and university.

Like Hall (2012) we witness and provide evidence of an underlying shift in values taking place over time. This is a transition from solidarity between and across the generations, to greater individualism and individuation. Tendencies to preserve privacy and maintain social distance with neighbours were common. It was good to mind your own business. They did not often visit each others' small cramped homes.

But they did stand ready to provide practical assistance, especially in challenging times, such as birth, death and sickness. Socializing tended to be mainly among family members, especially for women and girls. With a rise in disposable incomes and more secure jobs people begin to concentrate their efforts on purchase of consumer durables - cars and television sets, and holidays and sometimes travel to visit distant kin. Wide differences are visible between the foresight and planning of the careful aspirers and the feckless and reckless spenders, who have more outlets for leisure spending and consumption. In addition to pubs and clubs there are Bingo Halls, Cinemas, betting, dog racing and pigeon fancying.

Recently Kynaston (2009) has accumulated a huge amount of evidence on the working classes in the forties and fifties Britain. He writes of the sense of community among neighbours, all struggling to survive, of the importance of the pub, the pictures (cinema), football and the popular radio programs and annual rituals, such as bonfire night. He notes the later coming of slum clearance, high rise flats and council houses, social mobility through secondary education and the spread of car ownership and the "ubiquitous coming of the box in the corner".

Another point clear from a number of working class studies in different places has been that people were not homogeneous. There was often acute status consciousness, with clear distinctions between families and people living in different streets or ends of streets. Moreover as we have seen, a pervasive commodity enjoyed by most, especially women, was gossip - which helped to map out assumed variations in statuses and cultural differences, apparent to observers between different categories of people from different places and generations.

A Note on the Textile Heritage

Since this account was written, between 1975 and 1999 quite a bit has been done to try to preserve the architectural and mechanical heritage of Burnley textile mills and their employees. Just after the conversations recorded here were written down in 1977, the Burnley Industrial Museum Action Committee was formed, to promote preservation of the town's historical legacy and a Weavers' Triangle heritage site was established in 1980. The area comprises industrial buildings from the nineteenth century near the Leeds Liverpool Canal. It is a designated conservation area of remaining cotton mills, weaving and spinning sheds and warehouses, where unloading from the canal occurred. The restoration project, begun in the nineteen eighties, was designed to protect mills from demolition, so that they could remain as reminders of the past heritage. In the nineteen nineties a project began to restore the engine house and chimney at Oak Mount Mill. And in the same period Trafalgar Mill was also refurbished and Burnley wharf restored. This work was completed in 2001. These are now listed as Ancient Monuments. Other ambitious ideas to build marinas and luxury accommodation, such as turning Slater Terrace weavers' cottages into a luxury canal side hotel, have met various degrees of success and opposition. In fact some buildings have had to be demolished because they were structurally unsound and dangerous, such as Clock Mill Tower.

There have been multi million pound plans proposed, revised and rejected. There have been unfortunate fires and bankruptcies and the Council has questioned the feasibility of various proposals.

In 2012 work began on a technical college and her Majesty the Queen Elizabeth II visited the Weavers' Triangle, as part of her Jubilee tour of the North West of England. According to her speech her Majesty was pleased to visit Slater Terrace at the Weavers' Triangle. Its buildings are reputed to form one of the finest surviving Victorian industrial landscapes in the country.

On the outskirts of Burnley, at Harle Syke, is the world's only surviving, operational nineteenth century, steam powered, cotton weaving shed, with 308 Victorian Lancashire Looms. The boiler house, engine and weaving shed appear much as they would a century ago - if less dusty and dirty!

A heritage lottery project by CAR Video Unit has made possible the development of a film, showing what it was like to work in the heyday and decline of cotton. As part of the project young trainees filmed former mill workers, - looking for insights into a textile industry, which was once an engine of national production. This project *Memories of the Mills* has been aired in 2011 and 2012 on Television Channel M.

The BBC also have a video clip on *Sanitation and the Working Poor in nineteenth century England,* which sets out to provide a vivid description of the squalor and poverty of families in Burnley during the 1860s. In addition BBC 2 Nation on Film has an archive on *When Cotton was King* (last updated 2004 Cotton Towns Blackburn and Darwen). This links the industry to the earlier development of the linen weaving industry and the growing of flax around Preston and the making of woolen cloth in Blackburn in the sixteenth century. It calls attention to the weaving of Fustian, a cloth with linen warp and cotton weft, which was the first cotton based cloth produced in Lancashire from the early seventeenth century. Linen production is recorded as having spread to Burnley by the late sixteenth century and Fustian production declined after the invention of Arkwright's water frame made it possible to spin strong warp thread out of cotton. Prince Charles first visited Burnley in 2005 and stated that the attempts to regenerate the town were remarkable. Clearly there were challenges observed, as he put the town and a project to help disadvantaged youth at the top of his Prince's Trust Charity priority list. In the public imagination, in recent years, one could suppose that Burnley has been more often remembered for football prowess and race riots (2001). Maybe in future more images and memoirs will be of its productive past, as an important site of Britain's textile heritage and a place where women have played such a key part in the industrial work place, as adept and experienced minders of complicated machines, - for a time at least major makers of the world's supply of cloth. Finally there is news of attempts in Lancashire to lead a resurgence in textile manufacturing in Britain. There is evidence of a few new small factories weaving a variety of novel textiles, including woven grass! Rumour has it that even large machines themselves will eventually be woven.

BIBLIOGRAPHY

Briscoe L. 1971 **The Textile and Clothing Industry of the United Kingdom**. Manchester University Press.

Burn D. (ed.) 1958 **The Structure of British Industry: A Symposium Vol. II.**- Cambridge University Press.

Bythell D. 1969 **The Handloom Weavers.** Cambridge University Press.

Green A. 1970 **The Effects and Repercussions of the More Looms Systems in Burnley during 1932**. Chorley College.

Hall D. 2012 **Working Lives: the Forgotten Voices of Britain's Working Class** Bantam Press.

Hopwood E. 1969 **The Lancashire Weavers' Story** Amalgamated Weavers Association, Manchester.

Jones O. 2011 **The Demonization of the Working Class** Verso.

Kynaston D. 2009 **Family Britain 1951- 57**. Bloomsbury .

Mathias P. 1969 **The First Industrial Nation 1700-1914.** Methuen and Co Ltd.

Miles C. 1968 **Lancashire Textiles: A Case Study of Industrial Change**. Cambridge University Press.

Moser C. A. and W. Scott 1961 **British Towns: A Statistical Study of their Special Economic Differences**. Centre of Urban Studies Report No.2, Oliver and Boyd.

Neff W. F. 1966 **Victorian Working Women: An Historical and Literary Study of Women in British Industries and Professions 1832-50.** Frank Cass and Co. Ltd.

Nodal J. H. and Milner G. 1875 **A Glossary of the Lancashire Dialect.** Manchester.

Robson R. 1957 **The Cotton Industry in Britain**. London Macmillan and Co. Ltd.

Seabrook J. 1971 **City Close Up.** Allen Lane.

Smelser N.J. 1959 **Social Change in the Industrial Revolution: An Application of Theory to the Lancashire Cotton Industry**. Routledge and Kegan Paul.

Witham E.1940 **The Differentiation of the Cotton Industry Within The Weaving District of North East Lancashire**. Edge Hill.

Industrial Reports 1948 **A Survey Report of the Weaving Area of Lancashire.** Lancashire Industrial Development Association. Industrial Report No.2.

Industrial Report 1967 **The Decline of the Cotton and Coal Mining Industries in Lancashire.** Lancashire and Merseyside Industrial Development Association. Manchester.

CENSUSES H. M. S. O. LONDON.

1921 England and Wales: Preliminary Report.

1921 England and Wales: County of Lancashire.

1931 England and Wales: Preliminary Report.

1951 England and Wales: County Report, Lancashire.

1951 England and Wales: Occupational Tables.

1961 England and Wales: Migration Tables.

1971 England and Wales: County Report, Lancashire.

This book tells stories of textile workers, recorded in Burnley, in the heart of the traditional cotton industry in Lancashire, in 1975.

It was a time when the industry was going through final transformation before its virtual demise.

Married women still formed the backbone of the labour force of weavers.

They had not worn clogs and shawls for several decades.

What did they think of themselves and their jobs?

How did they manage to combine their several roles as mothers, weavers, kin and wives?

How did women and men relate to each other on the shed floor and in their homes.

What did they value? What did they gossip about?

What made them proud or anxious?

What were they striving for and aspiring to?

Gossip and tales from the factory floor give an inside view of their lives. And stories stretching in time over more than half a century indicate the rapidity and extent of socio-cultural and economic changes taking place during the period.

www.ingramcontent.com/pod-product-compliance
Lightning Source LLC
Chambersburg PA
CBHW050111280326
41933CB00010B/1048